AF697619

HOME POOL

Home Pool

The Fight to Save the ATLANTIC SALMON

by Philip Lee

GOOSE LANE EDITIONS and
NEW BRUNSWICK PUBLISHING COMPANY, LIMITED

Published by Goose Lane Editions and New Brunswick Publishing Company, Limited, with the assistance of J.D. Irving Limited, the Canada Council, the Department of Canadian Heritage and the New Brunswick Department of Municipalities, Culture and Housing, 1996.

Edited by Laurel Boone.
Cover photograph by Gilbert van Ryckevorsel, Aqua Marine Photography Limited.
Book design by Julie Scriver.
Printed in Canada by Unipress Limited.
10 9 8 7 6 5 4 3 2

Canadian Cataloguing in Publication Data

Lee, Philip, 1963-
Home pool

Includes bibliographical references and index.
ISBN 0-86492-200-0 (pbk.) — 0-86492-202-7 (bound)

1. Atlantic salmon fishing — New Brunswick.
2. Fishery conservation — New Brunswick.
3. Fishery management — New Brunswick.
I. Title.

SH685.L43 1996 333.95'6 C96-950159-5

Goose Lane Editions
469 King Street
Fredericton, NB
CANADA E3B 1E5

Contents

Preface

Our company, J.D. Irving, Limited, is pleased to sponsor the publication of *Home Pool: The Fight to Save the Atlantic Salmon* by New Brunswick author Philip Lee.

His series of articles, which were adapted for this book, was widely acclaimed when it first appeared in the *Telegraph Journal*. Philip Lee's skill in researching and telling the story of the great Atlantic salmon in our province's rivers immediately spoke to his readers. His work won praise and awards from journalism colleagues, from salmon fishermen and from conservationists.

Caring for and protecting our natural resources are responsibilities all of us must share. *Home Pool* will bring the story of the great Atlantic salmon to a wider audience and, we hope, move them to action.

JAMES D. IRVING
July 1996

For my father, who showed me the magic in rivers.

Foreword

In nature," wrote Robert Ingersoll, "there are neither rewards nor punishments — there are only consequences." As animals, people live as much in nature as birds and fishes; only we, amongst the creatures of the earth, can know the inseparable connection that exists between present act and distant consequence. A common sparrow has more than 2,000 feathers to maintain. It will spend 20 minutes preening them after each bath, waterproofing them with oil from a special gland, without knowing that its survival depends upon this diligent ritual. It will rid its body of parasites without knowing that a groomed bird supports only one-tenth as many lice as an unkempt bird. Knowing not a single thing about tomorrow, beavers will maintain a dam across dynastic generations and eagles have kept a single eyrie in good repair for as long as 40 years. Ignorant of the chemical properties of stickiness, a solitary garden spider will replace its web every day or two. And an Atlantic salmon — though it be the last one left on Earth — will still set off down the River of Time for its essential rendezvous with the sea. As the Victorian poet Matthew Arnold observed of all animals:

> Bounded by themselves, and unobservant
> In what state God's other creatures be,
> In their own tasks all their powers pouring,
> These attain the mighty lives you see.

In the end, existence means the proper taking care of things. For Robert Pirsig, in his classic *Zen and the Art of Motorcycle Maintenance*, proper care defines God better than love itself. Without disciplined and dedicated attention to the task at hand, human action must lead inexorably to dismal ends. In common usage, although not necessarily in theology or law, we correctly regard such delinquent action either as sinful or as criminal.

In nature, healthy animals meet their ordained responsibilities — self, home and territory. They do so without apology and without guilt. Sparrows aren't egotistic because they preen and beavers aren't materialistic because they build fine houses. Survival is one simple, obvious consequence of good hygiene and good shelter. This is not a modern revelation, although it is appalling how often and how completely people have forgotten it. When John Wesley surveyed the grotesque filthiness of eighteenth-century England and declared that cleanliness was next to godliness, he launched a moral revolution. Placing responsibility on the individual, Wesley turned practical morality into practical religion.

In this sense, Pirsig — more than 200 years later — preaches a distinctly Wesleyan sermon. People are responsible for the proper care of their own environments. People must act as their own environmental protection agencies. Moral conduct requires that we first fix, then maintain and, finally, improve whatever parts of our home-turf environments have gone wrong. (When people told Wesley that things were already as good as they could be made, he observed that "perfection can be improved.")

Harvard University professor James Q. Wilson has reduced this principle of moral conduct to a single rule. In his book *Thinking About Crime*, Wilson described an experiment conducted by a psychologist at Stanford University who had abandoned a car, licence plate removed and hood up, on a suburban street in affluent Palo Alto, California. He abandoned another car, comparably derelict, in a broken-down area of the Bronx. One week later, the Palo Alto car remained as he had left it; in the Bronx, people had stripped the car to its frame in 24 hours. He finally took a hammer to the windshield of the car in Palo Alto and abandoned it again; within hours, the car had been thoroughly stripped. The principle Wilson articulated was this: Anything that is broken and not quickly repaired subverts moral conduct and induces criminal acts.

Home Pool is a journalist's examination of the stripped salmon rivers of New Brunswick, now abandoned by the nominal owner, the Crown. The author, Philip Lee, is a colleague of mine at the *Telegraph Journal*, New Brunswick's provincial newspaper. (He is the editor of the *New Brunswick Reader*, a Saturday magazine produced by the *Telegraph Journal* and its partner newspaper, the *Saint John Times Globe*.) He began work on this project early in the summer of 1995, seeking to identify (1) what has been broken in the legendary watersheds of this province and (2) what the people of New Brunswick need do to repair the damage.

He travelled on the rivers of New Brunswick, and he talked with people who love them and who work to protect them. He went to Quebec; he went to Scotland and Iceland. In these places, he saw how other people conserve their salmon stocks. Four months later, beginning in late autumn, his front-page stories ran in the *Telegraph Journal* for 36 consecutive days as a series entitled *Watershed Down*.

In the months since its publication in the *Telegraph Journal*, the series has won many journalism awards, among them the Canadian Association of Journalists' prize for the best investigative writing in the country for 1995. Now, with the publication of *Home Pool*, the series has transcended the normal one-day life span that befalls the work of all newspaper writers. *Home Pool* justifies the definition of good journalism as literature written on the run.

Oscar Wilde once said, less kindly, that the difference between journalism and literature was this — that journalism was unreadable and that literature was unread. In this comment, Wilde affirmed the earlier assertion of Lord Bryce, who had said that the formula-writing found in most newspapers commanded people to forget it the moment they had read it. He lamented journalism's "constant succession of transient ideas, none of them impressively stated, though many of them startlingly stated, all of them flitting swiftly past the mental sight as trees flit past the eyes when one looks out the window of a train." This rebuke off his chest, Lord Bryce nevertheless observed that a single edition of a well-written newspaper was worth more than Thucydides' *History of the Peloponnesian War.* And Thomas Arnold, father of the poet, wrote in his journal: "I have just been looking over a newspaper, one of the most painful and solemn studies in the world when read thoughtfully."

Why this variance in judgement? In its most fundamental purpose, a newspaper is a daily journal of moral conduct. Newspaper editors and newspaper writers must be, as the American writer Ambrose Bierce put it, "severely virtuous censors." The censor in Roman context was a person who rendered judgements on public spending; the Latin word itself means merely "to estimate." As Bierce's phrase succinctly suggests, journalists must serve as auditors of moral conduct who can, again in Matthew Arnold's words:

> Take the suffering human race
> And read each wound and weakness clear
> And stick a finger on the place
> And say, "Thou ailest here, and here."

In this sense, newspaper stories are parables drawn from real life. The best of them are memorable. And literature, by definition, is simply memorable writing. It is not by style alone that the written word becomes literature; nor is it by content alone. Literature is writing — whether found in a daily newspaper or in a good book or in a letter to a friend — that transfers from page to imagination an imprint, an impression, that endures. At its highest level, writing is storytelling that creates authentic, permanent pictures in the mind. It is a biological form of "television." Thus the relevance of Lord Bryce's dismissal of writing that flits "swiftly past the mental sight." One reads — and simultaneously sees, feels, senses. One files in one's mental data bank, for instant playback, images worthy of preservation.

Are we at risk here of taking ourselves too seriously? Do newspaper writers not exist merely to fill the spaces between the ads? Journalists themselves sometimes think so. Some journalists understand the importance of the written word and the stories they write; many don't. Readers, I think, always do. The written word remains the primary influence on human action in the modern world. And newspapers remain — beyond all dispute — the primary forum for the written word. Newspapers remain the literature of the people.

"There is nothing more charged with the detonators of dreams and actions," writes George Steiner in *The Archives of Eden*, "than the written word." And he defines the authentic writer, the authentic editor, as "the obsessed servant of the text."

There is great argument among newspaper people these days over this question. Daunted by Electronic Age competitors, embarrassed by the idiosyncratic Industrial Age character of their business (ink on paper, delivery trucks), alarmed by a slow loss of readers, some newspapers are now more obsessed with the electronic delivery of data-bank information than with the writing of memorable stories. I believe that these newspapers are making a tragic mistake. Information, as a commodity, is valuable. I welcome its bulk sale by the gigabyte. Yet no mass of facts can replace the inherent truth and beauty found in the merging of right words.

More than anything, though, newspapers are embarrassed by their association with moralism, which places them in the awkward position of children who don't want to be seen walking in public with their parents. I am in the midst of reading a lecture given by a celebrated American editor, who declares: "There isn't any one way. There isn't any one truth. There isn't any one wisdom." As self-evident as these statements may be in an abstract philosophical way, they reveal the shame that burdens contemporary editors who no longer believe in the implicit instruction of their own Page One headlines. For almost every headline, in almost every newspaper, makes a public statement of some kind on moral conduct.

What this editor might better have said was this: "There isn't any one political party. There isn't any one political ideology. There isn't any one editorial-page commentator." These assertions would help restrain journalism from its stalking of the political process, of government and of the State. Government, as everyone knows, is inherently amoral and often immoral. It is this intrinsic characteristic of government, after all, that produced journalism's historic, original mission — its obligation to serve as the watchdog of government. Looking back, we can now see that the watchdog hasn't done its job very well, notwithstanding all the barking it has done.

Journalism's compromises with government arose from the intimate relationship that came quickly to exist between the two estates. Sir John A. Macdonald used to mail news leaks to the *Toronto Leader*, complete with his instructions on how prominently the stories were to be published. By 1919, H.F. Gadsby, writing in *Saturday Night*, could declare that Canadian journalists had become, across the country, "honk-honking propagandists . . . as comfortable [in their alliance with government] as a boll weevil in a cotton pod."

When journalists did crusade, they mostly sought more government. Although many of the reforms that they demanded were noble, they consistently confused reform with legislation. They became evangelists for the State, preaching government as a secular theology. Now governments everywhere are confessing their incompetence, conceding their waste, acknowledging their limitations. The statist sermons preached by many journalists are no longer universally credible. Decline in readership has tracked this disenchantment.

The moralist, however, does not seek salvation. The moralist seeks survival. The moralist wants to know what works, wants to know what's broken, wants to know who holds the responsibility for fixing it. In her mind-cleansing treatise *Systems of Survival*, Jane Jacobs makes this point brilliantly. In this book, nominated for the Governor-General's Award for 1992, Jacobs examines the moral systems that sustain civilization. She identifies a "guardian syndrome" that governs politics and a "commercial syndrome" that governs business and trade. Each of these equally necessary syndromes, she says, operates on distinctly different — and contradictory — moral dynamics. Moral rules among the guardians, she says, require that they shun trade, adhere to tradition, respect hierarchy and use deceit for the sake of the task. Moral rules among people engaged in business, she says, require that they shun force, reach voluntary agreements, be honest, be efficient and compete. The moral confusion of our time, she says, reflects a pervasive mingling — and a mutual corruption — of these diametrically opposed moral systems.

What makes Jane Jacobs especially relevant here is that she relied extensively upon newspaper stories for her evidence. Using news stories that appeared in the *Globe and Mail*, the *New York Times* and the *Wall Street Journal*, she illuminated the moral implications of this ancient aphorism: "All things have two handles; beware the wrong one." And she observed, in passing, that newspapers have a difficult job precisely because they are themselves both guardians and traders.

To restore their diminished relevance, journalists must come to understand again the moral dynamic of their work. They must learn to distinguish what works from what doesn't work. On this point, William Thorsell, editor-in-chief of the *Globe and Mail*, made germane observations in a lecture he gave a few years ago:

> The five Ws of reporting (what, who, where, when and why) are heavy on description. I believe that the five Ws must be supplemented by the capital-H of How and the capital-SW of So What. How do things work? How do things happen? Why does it matter? Print reporting must go beyond detailed descriptions of what happened to what we might call "adult education."
>
> The classic definition of news can be summarized as 'what went wrong yesterday.' This traditional definition of news has made very little room for 'what went right yesterday.' Surely the uncovering of things that go right should be considered a valid form of investigative journalism.
>
> When we speak so reverentially of investigative journalism, we mean uncovering something that has gone wrong — the wronger, the better — growing out of the classic but intellectually careless aphorism that it is the role of journalists 'to comfort the afflicted and to afflict the comfortable.' Why, in principle, would we want to afflict the comfortable?

It's a good question. The "virtuous censor" is not employed as a critic merely to condemn. Why are newspapers so reluctant to report success? Why are they so eager to ignore it, to dismiss it — and, often, to attack it?

Philip Lee's stories transcend the formula-writing of the traditional investigative newspaper series. He has vividly documented what went wrong under federal and provincial guardianship of New Brunswick's rivers — a long period during which hundreds of politicians, bureaucrats and scientists were amply paid to preside over the demise of the Atlantic salmon and the destruction of their rivers. He has spent much more time, however, documenting — in terms of responsible fishery management — what works. His conclusions, which are his own, do not surprise me, for they are consistent with nature itself and with moral responsibility — that territory and turf are best managed by the people who are dependent upon them, who are induced (by self-interest, by simple respect and by love) to tend them for the long term.

More than this, however, I carry from Philip Lee's series of articles a number of indelible mental images. Among these are the people of New Brunswick's rivers — the energetic, often eccentric "friends of the rivers." Foremost, though, are images of the salmon themselves. In these images, salmon circle in pools of pristine water, leap through the surface, form arcs in the air and hover there in split-second majesty. These images answer the question: What does the Atlantic salmon need for its survival? Nothing complicated at all: clean waters in unobstructed rivers — rivers kept forever, in Matthew Arnold's phrase, "as new-born, clear-flowing streams." Who should bring back the clean waters, the clear passages? People who care enough to do the job.

I was watching a video movie the other day, a history of the discovery of King Tut's tomb. It turns out that this great archaeological site, dug up only 75 years ago, is now crumbling away. Here was a piece of human heritage that had survived for more than 3,000 years — and the Egyptian government has allowed it (and many other similar sites) to approach ruin in only a few decades, an act of criminal neglect. Exploiting these ancient sites, Egypt has sold them off by the glance to tourists whose touch, step and breath have combined to destroy the very objects of their fascination.

The Atlantic salmon is as unique as that tomb. It, however, has survived for millions of years. It is one of Nature's most cherished species, famous for its epic journey to spawn, celebrated for its legendary will to survive. In the span of mere decades, it is as imperilled as Tut's tomb. In comparison, the destruction of Egyptian artefacts is mere delinquency. The destruction of the Atlantic salmon — this is inherently criminal.

In *Home Pool*, you will see an epochal murder mystery taking place in front of you, the indifferent torture and the wanton destruction of a noble creature. And you will meet the people who can avert the ultimate tragedy. Conscience calls out that they be permitted to try. Moral conduct commands it. In this book, Philip Lee — himself an intimate friend of the rivers and the creatures he would save — leads us to a moment of decision. Whatever choice the people of this province make, I am honoured to have been associated with the purpose and the work of this fine writer, with his colleagues in this project and with the censorian newspaper that produced them.

C.S. Lewis said that people read to be not alone. Gazing now at the social and environmental destruction around me, I turn to the company of the great Victorians, the last generation of writers who understood the inseparable connection between moral conduct and human survival. "Nature," said Matthew Arnold, "forgives no debt." We can choose only to pay it back — or to not pay it back. Each moment, each human act, carries with it a consequence for what Arnold described as "the quiet watershed / Whence, equally, the seas of life and death are fed."

To make the right choice, we must regain a sense of eternal things, of the night winds, of Arnold's "murmurs and scents of the infinite sea." Among these eternal things must surely be the mighty Atlantic salmon, its oceanic migration done, homeward-bound to the pool of its birth through the newborn rivers of New Brunswick.

NEIL REYNOLDS
Publisher and Editor-in-Chief
Telegraph Journal

And wherever the river goes every living creature which swarms will live, and there will be very many fish; for this water goes there, that the waters of the sea may become fresh; so everything will live where the river goes.

— *Ezekiel 47:9*

Always the rivers run, and always there will be great ships upon the tide, always great horns are baying at the harbour's mouth, and in the night a thousand men have died while the river, always the river, the dark eternal river, full of strange secret time, washing the city's stains away, thickened and darkened by its dumpings, is flowing by us, by us to the sea.

— Thomas Wolfe, *Of Time and the River*

CHAPTER ONE

Headwaters

Miramichi Morning

The water is the eldest daughter of Creation,
the Element upon which the Spirit of God did first move.
— Izaak Walton

On a blue morning between the banks of the Main Southwest Miramichi River, water ripples around a bend, reflects shimmering yellow light, rolls over round stones and pushes silently into a deep salmon pool. A ribbon of water under a big Miramichi sky. Sweet river music tightens the heart. On a Miramichi morning, gusts of wind ride along the current and history swirls down from the tree tops. The spirit of creation moves upon these waters.

The story of the wild Atlantic salmon begins here and moves thousands of kilometres into the ocean, then returns to fresh water, up big rivers and tiny streams, deep into the heart of the New Brunswick wilderness. *Home Pool* is a story about Atlantic salmon, their destruction, and our last chance to make amends. We have driven the Atlantic salmon, the greatest of all freshwater fish, to the edge of extinction. On a Miramichi morning, as I wade beside the bank with currents swirling around my feet, I am aware that thousands of big salmon rest in this river's deep pools. I am also aware that our actions during the next decade will decide the future of these salmon and their progeny and the future of our rivers.

The sun works hard to burn the night chill out of the air as fishing guide Renate Bullock wades along the edge of the river, casting her way down through the Elbow Pool, a narrow run that bends in the shape of a crooked arm around a rocky island. This is one of three salmon pools Renate and her husband Fred own just upriver from their red log home in Boiestown. Renate is lean and compact, a sure-footed athlete in green hip waders, an artist on the water as she throws a long line in an effortless motion. She tells me about her waters as she moves steadily along the side of the pool, expecting a salmon to rise to her fly on every cast.

"This stretch here, there's just something almost mystical about it, the way the water comes down over those rocks." On a gravel bar downstream, a doe and fawn

Renate Bullock works her way down through the Elbow Pool upstream from her home in Boiestown, New Brunswick, on the Main Southwest Miramichi River. PHILIP LEE

emerge from a stand of alders and wade into the shallows to drink from the river. "This is one of our blessings, to be able to watch nature at its best and then to have fish in your pool," she murmurs. "You can't help but appreciate the creation of it all." The deer look upriver toward us and then disappear silently back into the trees.

"Isn't that beautiful? If people were only able to stop and realize how beautiful nature is and then do something about holding onto it. Many times I have come up to this pool or one of our other pools and thanked God for the opportunity that I've been given. Most people are blessed one way or another in their lifetime, and I feel this is my blessing. I have never ever lost that feeling of awe, of respect and pride, and I've worked very hard to keep it. It's a deep-rooted pride. I love this water."

Renate guides anglers for Boiestown outfitter Vincent Swazey, her neighbour, close friend and fishing partner. Swazey leases the three salmon pools from the Bullocks. He lives in a riverside bungalow, a stone's throw from his family's white Miramichi homestead. His brother lives next door. On the family property, he operates his Tuckaway Lodge and guides salmon anglers. Swazey learned to pole a canoe and cast a fly line the way most of us learn to walk and talk — he's good at it, always has been, but he doesn't remember learning how. "The river means everything in my life," says Swazey, a vice-president of the Miramichi Salmon Association and one of the leaders of the river's conservation movement. "I spend a lot of time working for the river, and I can only try to put back as much enjoyment as I've taken out."

During the spring and summer of 1995, I walked the banks of great salmon rivers in New Brunswick, Quebec, Scotland and Iceland. Everywhere I went, from New Brunswick's Miramichi and Restigouche, to Quebec's Grande Cascapédia, Scotland's Royal Dee and Iceland's Nordura, I met friends of the rivers, defenders of salmon waters, people like Renate Bullock and Vincent Swazey, whose lives run with river currents.

Vince Swazey poles his canoe on the Main Southwest Miramichi River. THOMAS R. PERO

But these friends of the rivers haven't been able to do enough. While they have been protecting their waters, fishermen have been recklessly slaughtering fish in rivers and in the ocean; hydroelectric companies have been damming salmon rivers, blocking fish passage to and from the spawning grounds; industries have been polluting the water, air and land, upsetting nature's balance. The survival of the Atlantic salmon, and many other living things, has been threatened.

*

The story of the wild salmon is more than one million years old. Salmon were swimming up rivers to spawn long before humans began to inhabit the Earth. Cave dwellers who lived in Europe 15,000 years ago painted pictures of Atlantic salmon on their walls and carved the fish's shape into pieces of antler. In the 10,000 years since the glaciers retreated at the end of the last Ice Age, the salmon we see today in our rivers have evolved. Each river has developed its own distinct genetic stock of salmon, adapted precisely to the conditions in its home waters. No two rivers are the same; no two rivers have exactly the same genetic type of salmon. The Miramichi River system has more than a dozen distinct genetic stocks and many substocks, which have adapted to their particular streams. In the last 150 years, hundreds of salmon rivers have been destroyed, their unique genetic stocks of fish lost forever.

*

Every year, scientists at the International Council for the Exploration of the Sea, an organization of leading scientists from Europe and North America, estimate the number of salmon feeding in the sea. Their findings are alarming. The number of Atlantic salmon in the ocean has reached an all-time low. If these scientists' estimates are correct (and often their estimates of salmon numbers have been too high), and if the downward trend they have recorded continues, wild Atlantic salmon could be extinct in many rivers in 10 years. In the late 1960s, scientists estimated there were about 850,000 large salmon from North American rivers feeding in the North Atlantic. By 1995, they estimated that there were only about 150,000, less than the bare minimum of 200,000 large spawners scientists believe we need to seed our rivers. "The graph is heading down," says Dr. Alex Bielak, a salmon biologist, conservationist and former director of fisheries management for New Brunswick. "There have been periods of cyclical abundance, and even with that cyclical abundance, this is the lowest trough there's ever been."

The decline has continued even in the face of a dramatic reduction in the yearly Canadian kill of salmon. Since the early 1970s, the federal Department of Fisheries and Oceans, which is responsible for salmon conservation, has been gradually phasing out ocean net fisheries; Ottawa has spent more than $70-million buying out commercial Atlantic salmon licences, and so some of the fish that would have been caught by ocean nets are now migrating up our rivers.

"We've come to the end of the line in terms of closing the fisheries," says Dr. John A. Ritter, a salmon biologist for the Department of Fisheries and Oceans. Ritter, who has offices in Halifax and Moncton, heads the government's Maritime salmon research division. "We can't close any more and now we're beginning to see that the decline is still continuing, although one hopes that it's bottomed out by now. We didn't get the benefits from those closures [of net fisheries] that we should have. If you go back and look at how many large salmon were taken in the Miramichi alone back in the 1960s when the [net] fishery was operating, we are not getting those large salmon there today." Too many of the salmon that should have returned to our rivers after the nets were lifted are dying at sea.

*

"The salmon are a test of a healthy environment, a lesson in environmental needs," wrote Canadian angler and naturalist Roderick Haig-Brown in *The Salmon*. "Their abundant presence on the spawning beds is a lesson of hope, a reassurance that all is still well with water and land, a lesson of deep importance for the future of man. If there is ever a time when the salmon no longer return, man will know he has failed again and moved one stage nearer to his own final disappearance."

Salmon conservationist Dr. Wilfred Carter, president emeritus of the Atlantic Salmon Federation, an international conservation organization based in St. Andrews, New Brunswick, speaks of the decline of the Atlantic salmon as a symbol of our misuse of renewable resources. He says we have only to look around us to see other examples of human destruction and greed: "We are eating our way down the food chain. We've destroyed the northern cod and the North Sea herring — fish that once numbered in the hundreds of millions, fish so plentiful their extinction couldn't have been imagined 50 years ago. Fishing fleets have overfished capelin and other tiny fish, the food sources of cod, salmon and other larger fish. Some of these tiny food fishes have been harvested to burn in generators to produce electricity. Now we are going after seaweed and marine plant life. On land, we've destroyed forests and allowed rich soil to be flushed into rivers. We've dammed rivers. We've allowed poisons to flush into rivers, lakes and the sea.

"As people who want to go on living and enjoying the spiritual and material blessings which are available on this planet, we have an interest in the Atlantic salmon that transcends the simple urgency of preventing the disappearance of a magnificent fish which has brought profit and pleasure to millions. If we act successfully to save the salmon and to clean up rivers it has abandoned, we will be making it possible for other things to go on living too, and the life of each one of us will be enriched, even if we never even see an Atlantic salmon. We now know only too well the price which will be exacted eventually if we continue this relentless confrontation with nature. When we see conditions restored in rivers which enable the salmon to return again, we will have reached a proper understanding and respect for the Earth's life-support systems. There will then be hope that man himself can survive."

We have learned from decades of government management of our rivers that when we compromise on clean water we end up with dirty water, when we compromise and build dams we destroy salmon runs, when we compromise and overharvest fish we drive stocks into extinction. I am proposing a transfer of responsibility from governments to the friends of the rivers — people who will not compromise on clean, clear, free-flowing waters.

*

The Atlantic salmon miracle begins when females deposit tiny eggs in gravel nests as the water cools in the fall. Males, competing for positions nearby, rush in to fertilize the eggs which, covered over with gravel, lie still and quiet during the long

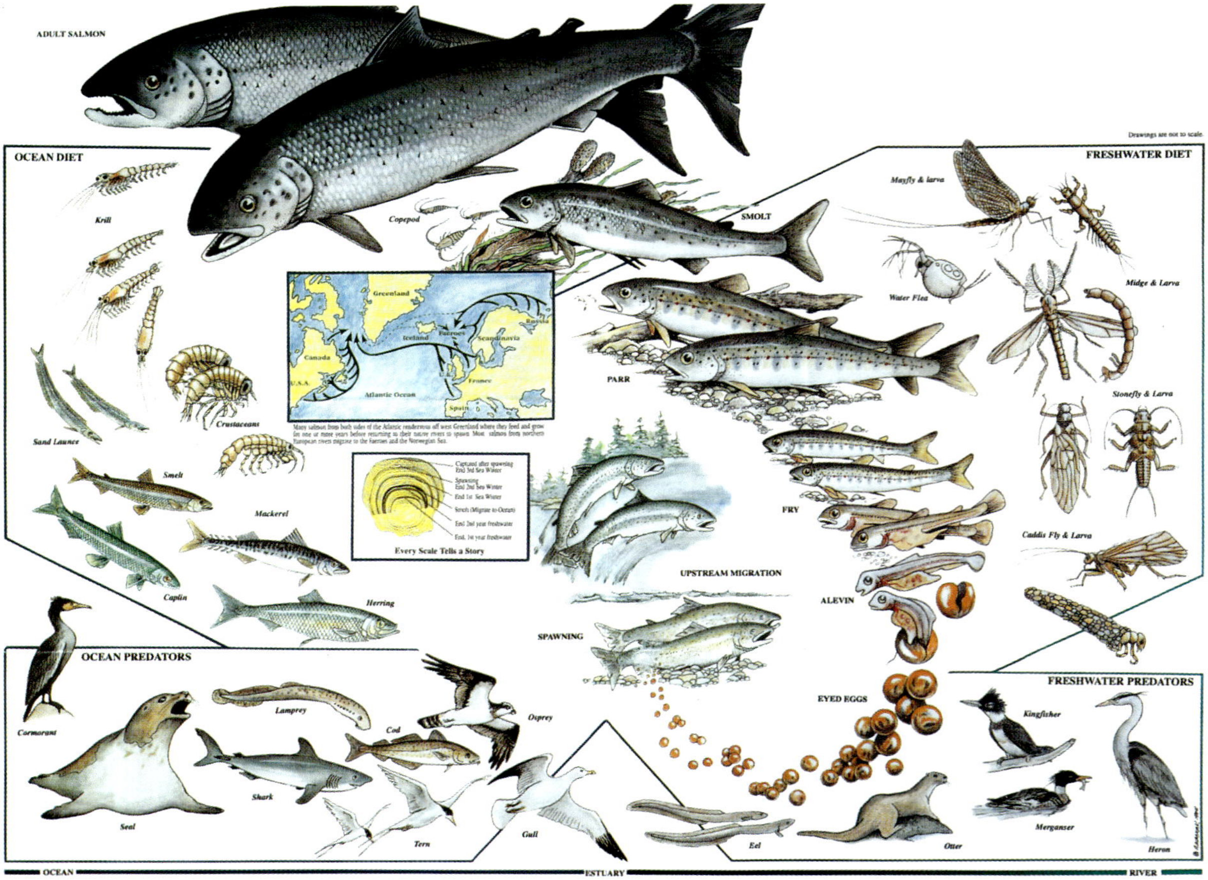

The miracle begins in fresh water: the life cycle of the Atlantic Salmon.
ART MacKAY STUDIOS

winter months. In early spring, as river water warms, tiny fish tails break through the skin of the eggs and baby salmon called alevins are born. The alevins are nourished by yolk sacs which hang from their bellies. Until the yolk sacs are absorbed, the alevins stay hidden in the gravel, gathering strength for the time when they will begin to feed themselves.

Baby salmon about two centimetres long, now called fry, emerge from their hiding places in the gravel and begin to eat tiny organisms in the river. When they double in size, they become known as fingerlings. When they grow to be about eight centimetres long, they are called parr and develop brown backs with black and red spots on their sides. In most rivers, only 10 per cent of eggs survive to become parr. The parr grow slowly, feeding on insects and insect larvae. Parr remain in fresh water between two and seven years, depending on the conditions in the river. They are wonderful athletes for their size — a parr 10 centimetres long can jump almost a metre out of the water to drag an insect down into the river for its meal.

In April, as alevins begin to hatch in the gravel beds, parr prepare for their new life in the sea. Their red spots disappear and their sides take on the new silver colouring that will help protect them from predators in the sea. Their tails lengthen and become more forked. These young fish, called smolt, look like miniature adult salmon. They swim downstream with the current and ride in and out of the mouth of the river on the tides, adjusting to the salt water as they prepare to head off to feeding grounds in the North Atlantic.

Atlantic salmon often leap high into the air in fresh water.
ATLANTIC SALMON FEDERATION

By mid-summer, young salmon from New Brunswick rivers have travelled 1,000 kilometres to feed off the coast of Newfoundland. By fall, some have moved another 1,000 kilometres into the Labrador Sea. They then move east, feeding along the coasts of Canada and Greenland. They grow rapidly, gorging on shrimp, capelin, herring and sand eels. After one year at sea, a salmon can weigh up to six pounds; after two years, 20 pounds; and after three years, as much as 40 pounds. Some salmon, called grilse, return to their native rivers to spawn after only one winter at sea. Fish that remain in the sea more than one winter are called multi-sea-winter salmon.

These fish are international travellers. Salmon from Europe and North America meet on the feeding grounds off the southwest coast of Greenland. This winter feeding party includes salmon native to Canada, the United States, Great Britain, Norway, Sweden, France, Iceland and Russia. The same sort of party takes place near the Faroe Islands between Scotland and Greenland and at other locations in the North Atlantic that remain the salmon's secret.

When salmon are feeding at sea, the maturation of their sexual organs triggers an instinct to go home. Then their amazing journey begins. Guided by the sun, the stars and electrical impulses from the Earth's gravitational field, they travel thousands of kilometres back to the coastal area of their home rivers. Then they rely on their ability to detect minute chemical smells in the water to find their way home.

"They swim through the outflow of river after river after river as they're swimming along the coast," Carter says. "And when they swim through the outflow of their own natal river, they recognize it and instantly turn into that river and head up it. Not only do they recognize their own river, but they'll move on up the river and reject tributary after tributary until they come to the tributary that they recognize. They'll home up that tributary and very often go back to the very pool, the very spot, they left two or three years earlier."

When they reach the outflow of their home river, Atlantic salmon often leap into the air before swimming upstream, as if to celebrate their return home. Fresh from the ocean, a spawning salmon is in perfect physical condition, like a young, finely tuned athlete. But as the fish move upstream they begin to age, and their internal organs begin to deteriorate.

When the mature salmon reach fresh water, they stop feeding. This is nature's way of protecting the species, keeping the mature salmon from gorging on parr and fry in the rivers. They move at night and rest during the day in the same pools where salmon have rested for thousands of years. From time to time, they leap out of the river for no apparent reason, perhaps practicing for their ascent of falls that block their passage to the spawning grounds; some jump four metres in the air to make it over waterfalls.

In late October, males and females pair up and move over the gravel spawning beds. The male drives away other males and sexually mature male parr (called precocious parr) and chases away eels and trout that would eat salmon eggs. The female digs a nest (called a redd) and deposits eggs, and the male fertilizes them with milt. Sometimes they rest in a nearby pool, but they return again and again to the redd until all the female's eggs are gone: up to 750 eggs for every pound of body weight.

By the time the salmon have pushed their way upriver to spawn, they have aged so much they are near death. After spawning, about 90 per cent of them die and fall to the bottom of the pools; their remains are swept out of the rivers by the spring freshet. But 10 per cent survive and either make it back out to sea in the fall or spend the winter under the ice, staggering back to the ocean in the spring.

"As spent spawners returning to the sea, they will be gaunt caricatures of the sleek, firm fish of the earlier season," writes renowned American fly fisherman and author Lee Wulff in *The Atlantic Salmon*. "Yet even as they ready for the trip back to the sea, while still in fresh water, their sides and bellies will suddenly brighten again to silver. Before they reach the sea and start their ravenous recuperation, Nature touches them with her magic and makes them ready." The fish that make it back to the ocean are rejuvenated, their internal organs and muscles restored. They go back to the feeding grounds, then return to spawn again. Some Atlantic salmon have been tracked returning to a river to spawn more than five times.

New Brunswick writer David Adams Richards, who grew up alongside the Miramichi River, wrote of the salmon's fierce drive to spawn in his novel, *The Coming of Winter*, the story of a Miramichi boy's coming of age. Richards could see in the salmon's journey a metaphor of the human condition.

> It was said that their will was so great. He knew it. The way they followed up the water coming in from the ocean to come up along the river. And fighting everything as they did, splashing up madly out of the water to make the jump. It was said they would lie at the bottom gathering into themselves all the strength they had and wait and wait. Then they would begin at a point and make the run, gathering, gathering something frantic inside to come up out of the water with the surge they had gathered. Once up they would topple over or topple back, continue onward or slant on their side at the bottom again. Those that failed would try once more, exhausted they would flick their tails up and out, never reaching, never quite reaching. Battered and bleeding in the attempt. Those that continued would spawn, would loosen their eggs or sperm into the soft mudbank of the stream, loosen in their progeny the life within their flesh. As if some great release to existence they were fated to give came with something destructive within them all. As if, he thought, something comical within it all.

*

"The concern I had for a long time was that not enough people cared," Wilfred Carter says. "There were a few people who cared about salmon and birds and wildlife, but the average person really didn't understand the importance that those things had in their lives. They didn't understand that for people to survive, you had to have a healthy environment. Healthy forests and healthy rivers and all of those things impacted on people's ability to survive. You really can't separate the fortunes of wildlife and the environment. You can't really separate those from the fortunes of people. If we really screw up everything badly, we are going to end up ultimately not being able to survive ourselves in the environment that we've created. My concern for a long time was that people didn't understand this. I'm not sure that people do yet, by and large."

Jack Fenety, the president emeritus of the Miramichi Salmon Association and the dean of salmon conservationists in New Brunswick, agrees with Carter. The former radio broadcaster has been fighting conservation battles tirelessly, on many fronts, for more than 40 years. He says the conservation movement has made great strides but hasn't advanced nearly far enough. "I was one of the early conservationists in an active sense. The people in those days, as is the case today, believed the water would always be clean and the trees would grow. The apathy is limitless. It's really tragic. It's appalling how few people will financially or actively or otherwise support conservation. They will pay lip service to it, but they just don't do anything."

*

In what way do we need to care? Miramichi writer Herb Curtis describes in religious terms one man's connection with the Dungarvon River, a tributary of the Miramichi. In the final pages of his novel *The Lone Angler*, he writes:

> Palidin was sitting on a rock. He stood and walked to the water's edge, bent down and washed his face in the Dungarvon. As he washed, the water fell from his beard, trickled down his skin. He flicked it from his fingers, watched it drop like rain on the river.
>
> "Like rain," he thought. "I'm making it rain! Rain . . . a truly holy experience . . . It dances on the water, maps grey windowpanes with tiny streams, rises, falls, flows, flows in wells and springs . . . sometimes whispering, sometimes murmuring . . . I suppose it . . . it fell on the hair of Jesus, ran down his face to rise again until it falls to touch me . . . to touch us all, here and now . . . a blessing.
>
> "The Dungarvon . . . The Dungarvon River! Pure, clean, untouched, unspoiled . . . please never desert us!"
>
> Paladin waded waist-deep into the water, then submerged the rest of himself with a splash. He began to swim against the current, stroke after stroke after stroke, the current holding him stationary, though not in battle, perhaps in union — everything flowing; blood, water, molecules, thoughts, condensation, spirits . . . mutual respect, river and man, buddies like dew and rain, the universe and God . . . The fish gathered just to see.

The battle to defend the world's salmon rivers raises fundamental questions about how we live with nature, how we care for our natural resources, and how we govern our own lives. The friends of the rivers have found a way to live in a world of mutual respect. Where God and the spirit of creation move on the water. And salmon return from the sea to spawn in rivers as old as time.

A History of Wrecking Rivers

A few years, a little more overpopulation, a few fresh poaching devices and the salmon will be gone. He will become extinct.
— Charles Dickens

One hundred years after Charles Dickens lamented the demise of the salmon runs in the Thames, here in New Brunswick a Woodstock dentist, fly fisherman and amateur archaeologist was leading the fight to save the St. John River. Dr. George Frederick Clarke spent his days off on New Brunswick rivers, either fly fishing for Atlantic salmon or excavating ancient native campsites along the riverbanks. He would often confound his fishing guides by asking them to pull the canoe over to the bank, where he would put down his fly rod, drop to his knees and spend the day digging for artifacts. He knew the St. John River as well as anyone in New Brunswick, and he predicted that the new dam planned for Mactaquac would destroy one of the world's great Atlantic salmon rivers, which was already threatened by two major dams upstream.

He lost his fight to stop the Mactaquac Dam. And he was right — the dam destroyed the wild fish runs; only a shadow of the river's salmon runs has been maintained by a massive restocking effort at the base of the dam. The dam's headpond drowned the homes and history of 2,000 people and covered countless ancient campsites left behind by native fishermen who travelled on the river system. Clarke wrote 15 books and he never let our governments forget what they did to the St. John River.

"This destruction, we are told, is progress," he wrote in *Someone Before Us*. "A stagnant pond, 60 miles in length, has replaced the water wont to ripple over the bars and around the heads of the islands — the singing waters which, ever since the retreat of the glaciers, have washed the reflected stars and constellations as they moved across the bowl of the sky. Man is the great destroyer, the insatiable predator, but with all of his science he cannot create one blade of grass, nor the smallest flower, nor a tree, nor a bird, nor a waterfowl. True, he may reach the moon and the stars, but it's possible he may destroy them and this world as we know it."

England led the way in the progress that destroyed many Atlantic salmon rivers; the Industrial Revolution marked the beginning of the end. In the 1860s, Charles Dickens spoke out against the ruin of the salmon runs on the Thames, and, in 1968, Anthony Netboy described the demise of the world's salmon rivers in *The Atlantic Salmon: A Vanishing Species?* (the question mark was still appropriate). Netboy, a professor of English from Oregon who was obsessed with salmon, travelled to 10 European countries, studied thousands of historical and scientific documents, and produced a book that he hoped would be a rallying point for the preservation of the Atlantic salmon.

"The first true factory in England was built at Derby in 1718-22, a silk mill driven by water power from the Derwent, an excellent salmon stream," Netboy writes. He points out that the more than 350 mills operating on the Thames and its tributaries by 1868 doomed the Thames salmon, which needed to travel 65 miles above London to spawn. By the end of the century, salmon no longer migrated up the Thames. Then the Rhine, once the most productive salmon river in Europe, with fish spawning in clear water above Basel, Switzerland, was fouled by pollution and blocked by a series of hydroelectric dams. These obstructions, combined with heavy overfishing, ended the Rhine salmon runs in the 1940s. Salmon also deserted the Seine; the Gudenaa, Denmark's largest river; the Douro, Portugal's major river; the Mino, Lerez, Nalon and Nansa in Spain; and the Elbe and Weser in Germany.

The most shameful part of this story is that, when rivers were ruined and salmon runs wiped out, the people responsible for this destruction knew better. For hundreds of years people have known that for salmon to survive and spawn, they must migrate from the sea up clean, unobstructed rivers. According to Netboy, early salmon acts in England and Scotland were an attempt to keep river passages open and to protect young salmon: "The Magna Carta, the great charter of British liberty which King John was forced by the barons to sign at Runnymede in 1215, stipulated that all fish weirs placed by the king or his agents in the Thames and Medway 'and throughout all England except the sea shore' shall be removed and free passage for migratory fish assured. In 1285, Edward I set strict dates for the opening and closing of the salmon-fishing season to allow for spawning, and he prohibited the killing of fry."

Robert Williamson, Scotland's inspector of salmon and freshwater fisheries, says early salmon acts in Scotland showed a remarkable understanding of salmon biology. He cites an account from the 1520s, in which Scottish historian Hector Boece describes the life cycle of the salmon: "The salmon in harvest time cometh up into the small rivers, where the water is most shallow, and there the male and female rubbing their wombe one against another, they shed their spawne, which forthwith they cover with sand and gravell and so depart away. From henceforth they are gant and slender, and in appearance so leane that they appeare nought else but skin and bone and therefore worthily said to be growne out of use and season. The aforesaid spawne and milt being hidden in the sand (as you have heard) in the next spring doth yield great number of little frie but so nesh and tender for a long time, that till they come to be so great as a man's finger (if you catch anie of them) you shall perceive them to melt and their substance to dissolve and fade even as it were gellie or as ice laid forth against the sun. From henceforth they go to the sea, where within the space of 20 daies they grow to a marvellous greatness, and then returning againe toward the place of their generation, they shew a notable spectacle not unworthie to be considered."

The historian's only mistake was that the fish he saw coming back upriver had been feeding at sea for two or three years, not 20 days. In the mid-1600s, British naturalist Izaak Walton, in *The Compleat Angler*, described tagging of young salmon with a ribbon, tape or thread and observing how they returned to the same river where they were born.

*

In North America, salmon once filled all accessible rivers and streams from Ungava Bay to southern New England. The Atlantic salmon was an important food source for natives. Remains of ancient Indian salmon fishing weirs and campgrounds have been found at fishing spots throughout Quebec, Atlantic Canada and New England.

Atlantic salmon runs in United States rivers were the first in North America to be wiped out. The Connecticut River, flowing from the Connecticut Lakes along the border of New Hampshire and Vermont, through Massachusetts and Connecticut, was once one of North America's best salmon rivers. Salmon migrated 300 miles up this water system to spawn, but the runs were destroyed in the 1800s by a series of dams without fishways and by massive overfishing. The Merrimack in New Hampshire and Massachusetts was destroyed in the same way. Maine rivers were devastated by logging, which blocked the rivers and filled pools with debris. In Maine, the Penobscot, Kennebec, Androscoggin and St. Croix rivers all once had famous salmon runs. In New England today, salmon runs are a ghost of the great runs of the past, and a valiant restoration project on New England rivers that has cost more than $150 million has been discouraging. At the present time, there are actually more dams than Atlantic salmon in New England.

A salmon fishery was established in Newfoundland in 1708, and salmon were exported from Newfoundland and Labrador throughout the 1700s. In 1779, George Cartwright landed 12,400 salmon on the Eagle River in Labrador in less than a month. In 1764, William Davidson set up a salmon fishery on the banks of the Miramichi. The firm Robin, Piper and Company also fished the Miramichi, and the company's vessel, the *Hope*, transported salmon to Spain. In these years, more than a million and a half pounds of salmon were shipped out of the Miramichi system each year. There was a similar fishery on the famous Restigouche

River in northern New Brunswick. Fresh salmon packed in ice were shipped from Saint John by steamer to Boston, where they were repacked in ice and shipped farther south.

By the mid-1800s, salmon fisheries began to fail. R.W. Dunfield, in *The Atlantic Salmon in the History of North America*, describes a rapid decline in the amount of salmon exported from Atlantic Canada during this period. "The wholesale destruction of salmon and the subsequent catch decline was being described by commercial fishermen, anglers, fisheries overseers, and others in poignant terms, and even the casual observer could see that the fisheries in Upper Canada and southern New England were, by 1850, things of the past. Throughout the 1850s, the only region that appeared to sustain a salmon fishery in relatively undiminished proportions was Labrador, but it was predicted that even that remote area would soon be depleted of its stock, because of the rapacious fishing activity that was seasonally carried out and expanded." Dunfield quotes fishery officer James Hickson, who commented on the Bay of Chaleur fishery in 1883: "The steady decline in the catch of salmon on our coast indicates the end of this fishery at no distant day."

In the late 1800s, the federal government was involved in restocking tributaries of the Miramichi — the Little and Big Sevogle, the Bartholomew, the Cains, the Clearwater and the Dungarvon. But at the same time there was heavy netting at the mouths of all of New Brunswick's rivers.

Canadian rivers continued to be neglected and salmon runs destroyed. Waste from industrial plants, woodland operations, farms and private and municipal sewage systems spilled into rivers. Timber operations near rivers caused erosion, the filling of riverbeds, the destruction of natural spawning grounds and flash floods. And there were, of course, the dams. Netboy writes: "The 1890 report of the Nova Scotia Department of Fisheries disclosed that of 27 salmon rivers emptying into the Northumberland Strait, only four were unobstructed, and all others were blocked by mill dams without fishways!"

Pollution and overfishing drove the Atlantic salmon out of Lake Ontario and its tributaries by the late 1800s. Says Wilfred Carter: "An apparent illusion of limitless wealth from the sea stimulated an ever increasing search for new rivers to fish, without any alarm being sounded about the effect such plunder would have on future stocks; the inevitable result was not long coming as one after another the river output of salmon declined and then disappeared."

*

To make matters worse, for a decade New Brunswick carried out the world's largest fish-killing experiment. In 1952, the province began its battle against the spruce budworm, spraying 200,000 acres of woodland in the Upsalquitch watershed with the pesticide DDT. The following year, two million acres were sprayed in the Restigouche and Tobique watersheds. The next year, one million acres were sprayed in the northern Miramichi headwaters. The province continued spraying at a rate of about two million acres a year for 10 years.

In the first year of the program, Jack Fenety covered the spray program as a journalist for CFNB radio and the CBC in Fredericton. He was at first impressed by the job DDT did on the budworm. But he soon realized the spray was killing more than the budworm. "The following year it became evident that we were killing off all the trout, all the frogs, all the bees, everything," he says. "It was an appalling thing. It killed an awful lot of young salmon. It killed the squirrels, the birds. It killed everything."

"The effects of applying DDT in forests which shelter waters rich in fish life became quickly apparent," writes Netboy. "Fingerlings and small parr exposed to heavy and repeated dosages dropped by aeroplanes died by the millions. Older parr also suffered tremendous mortalities but were better able to survive, while adult salmon apparently were not affected. Fatalities occurred not only immediately after spraying but up to four months later."

In 1955, Dr. C.J. Kerswill, the scientist in charge of Atlantic salmon investigations at the Atlantic Biological Station in St. Andrews, described the devastating results of the 1954 spray program in an article published in the *Atlantic Salmon Journal*. "This had very serious direct effects on the young salmon, causing heavy mortality particularly among the fingerlings and small parr. In the sprayed streams most of the immature stages of all kinds of aquatic insects were killed also. These insects are the main food of young salmon, and some valuable kinds may not be plentiful again for years."

Two months after the spraying, the fingerlings were almost completely gone, small parr were reduced to one-sixth the 1953 level in the Northwest Miramichi and to one-half of the 1953 level on the Dungarvon, a Miramichi tributary. Other extensive observations on the effects of DDT were made by holding parr inside and outside the sprayed area and by patrolling for signs of dead and dying fish. These experiments showed heavy parr mortality both immediate and delayed. But, despite overwhelming scientific evidence that DDT was deadly to young salmon, the spray program continued until the early 1960s, when the province began to experiment with alternatives to DDT. Spraying for the budworm with another pesticide, fenitrothion, continued in New Brunswick until 1994.

*

Nothing has hurt salmon worldwide more than hydroelectric projects. A series of dams in the St. John River watershed has just about wiped out the salmon run in one of the world's great river systems. A small run is now sustained by the giant fish hatchery at the base of the Mactaquac Dam. "Next to the Miramichi, the St. John was the most productive salmon river in the world," says Bill Taylor, president of the Atlantic Salmon Federation. "It quite often had a run of 40,000 or 50,000 salmon, most of them large fish. Now, with this huge hatchery at Mactaquac, we're only getting 4,000 or 5,000 fish back. That's pathetic. It's a sin. But that's what happens when you put huge dams on salmon rivers. Dams and salmon just don't mix, regardless of whether you have state-of-the-art upstream and downstream fish passes. State-of-the-art is simply the best available technology at the time, and it just ain't good enough."

The federation is still spending money and energy fighting dam proposals on Canadian salmon rivers. In 1995, Taylor's organization was trying to stop a proposed dam two miles from the salt water on Newfoundland's Northwest River, a small, pristine, winding salmon river that flows through Terra Nova National Park. "To think that the provincial government is entertaining a proposal to put a hydro dam on that river is ludicrous," Taylor says. "In essence, you would block off 97 per cent of the river."

Before the Northwest battle, the federation fought to stop a hydro plan in the Moisie River watershed on Quebec's North Shore, one of the most famous salmon watersheds in the world. "How could the province of Quebec even think about such a thing?" Taylor asks. "Yes, we've got Mactaquac there to look at and say, 'How could we have ever been so stupid?' But we're just as stupid today."

*

For many years, the salmon was safe only when it migrated thousands of miles out into the North Atlantic. But in 1958, fisheries research vessels discovered that Atlantic salmon gathered at a common feeding ground off West Greenland. Fishermen immediately started killing them there in drift nets. Some of the nets were more than 20 kilometres long. In 1971, 750,000 salmon were harvested there. In the 1960s and 1970s, fishermen began exploiting similar salmon feeding grounds in the Norwegian Sea and around the Faroe Islands.

While salmon were being heavily fished off Greenland, the Canadian commercial Atlantic salmon fishery was being cut back. In 1972, the federal government closed parts of the New Brunswick commercial salmon net fishery and put restrictions on the Newfoundland commercial fishery, which was intercepting fish as they migrated back from their ocean feeding grounds to Canadian and United States rivers to spawn. In 1984, the Maritime fisheries were closed completely, and the Newfoundland fishery was placed under a five-year moratorium in 1992. But off the coast of Labrador, about 200 netsmen are still intercepting North American salmon. The Labrador commercial fishery took more than 50 tonnes of wild salmon in 1995; scientists from the International Council for the Exploration of the Sea have recommended that this fishery be closed. The net fishery has contributed to the decline of salmon stocks in Labrador rivers as well. Salmon stocks in remote Labrador rivers have been in steady decline for 15 years, to the point that the salmon population there is only one-quarter of what it was in the mid-1970s.

In 1995, fishermen off West Greenland harvested 71 tonnes of salmon, about 20,000 fish. The Greenland fishery has been drastically reduced from its peak of 2,700 tonnes — 745,000 salmon — in the early 1970s, but Atlantic salmon today are too scarce for this fishery to be operating at all. "All scientific evidence indicates salmon stocks off West Greenland are so dangerously low that the quota for the commercial fishery there should be zero," says Taylor. "This is madness."

CHAPTER TWO

The Watery Wastes

Guardian anglers

Mists dance above the water as the narrow cedar and canvas canoe skims down the river, pushed by the current and an outboard motor through narrow channels and over bubbling rapids. The Miramichi River, with its wide, sweeping turns below Doaktown, is almost deserted in the early morning light.

Marty Stewart, the chief guide at Old River Lodge, and his guest Don Senese, are motoring down to the Lower Pool to fish for the morning. Senese, from Victoria, B.C., is an experienced fly fisherman who has returned to New Brunswick for his second crack at fishing for Atlantic salmon. He is staying at a lodge owned by outfitters Alex and Vicki Mills, paying $385 a day, which includes fishing in private pools, meals, a room and a guide.

Stewart, one of the best guides and fishermen on the river, keeps a close eye on Senese, advising him about where to fish, tying the flies on his line, and making sure he doesn't wade out so far that he disturbs the salmon. At the same time, Stewart works around the pool, cutting out a piece of brush here, placing a rock there. It's like tending a garden. The river bank is spotless.

Upriver, Alex and Vicki Mills relax on the front porch of their lodge and look down the hill, where the Miramichi River meanders lazily past their wide front lawn and dock. Alex Mills is a lawyer, and he keeps his office at the lodge where he can watch the river through his window. He and Vicki run Old River Lodge, which houses eight fishing guests at a time. They have room to sleep 18 people, but they don't want to crowd their fine collection of salmon pools — they own or lease eight on the Main Southwest Miramichi.

Vicki Mills loves the river in every season, but especially in the winter months, when she goes alone to the lodge to tend the wood stove. "There's absolutely nobody else there. You look out across the river and it's all frozen solid and the wind's howling. Our whole life is run by the river, and it's kind of nice, because it's independent and you can't control it. It's always changing."

A river, like a living thing, moves in the spaces between stability and constant change. Friends of the rivers live their lives obsessed with the subtle nuances of

Guide Marty Stewart (left) watches Don Senese fish at Alex and Vicki Mills's Lower Pool on the Main Southwest Miramichi River. PHILIP LEE

their waters — the water level in the pools, the water temperatures, the position of channels, the lie of individual rocks, the holding patterns of fish, the rips of current, dark shadows and swirls.

Alex and Vicki Mills don't invest in stocks or bonds; they invest every dollar they have in salmon water, the most expensive real estate in New Brunswick. Owning salmon water is risky because you can lose your investment in one spring's ice run. Salmon pools move and disappear as the riverbed shifts and sends gravel and huge boulders downstream. The Home Pool in front of Old River Lodge (part of Alex Mills's initial investment in 1976) disappeared in the mid-1980s when an island migrated downstream and buried it.

There are different types of salmon pools. Some have a deep-water holding area where salmon rest, with a run of swift water upstream or downstream. Other pools are called bar pools, where a gravel bar with shallow water partly blocks the river. The salmon will splash over a bar only at night, so they hold down in the water in front of the bar during the day. Pools surrounded by a ledge of rock are the most valuable because they don't move.

Senese fishes through the morning along a long gravel bar. At the midway point of the pool, well back from the river's edge, is a bench marked with a sign: Private No Fishing. This long, swift, curving pool is the private property of Old River Lodge. The lodge's Lower Pool is bordered by two public pools — the famous Wasson's Bar downstream and Michael's Landing upstream. Like the rest of New Brunswick's salmon rivers, the Miramichi is a mixture of private and public waters. This unusual and controversial system of private ownership of salmon waters has been a powerful conservation force for more than 100 years. "There's fish left on these rivers because there's been a lot of conservation by private owners, a lot of wardening by private owners," says Vincent Swazey. "You try to take care of what's here, save what you have and manage it to the best of your ability, and hope it will always be here for your children and people in years to come."

There are more privately owned kilometres of river in New Brunswick than anywhere else in North America. Here, river ownership and exclusive fishing rights come in several different forms, all of which come from the fundamental concept of property rights.

A person who owns the land bordering a river in non-tidal water has what is called riparian rights, a series of clearly defined common-law rights attached to the ownership of the bank, or the *ripa*, of the river. Along with riparian rights, a person can own the riverbed to the midway point of the stream and can have the exclusive right to fish up to that point. A person can also buy the exclusive right of fishing, which can be separated from the land. Or, a person can actually own the riverbed, an entire pool from bank to bank, and therefore the exclusive right of fishing. These pools, the best of them, sell for millions of dollars.

Marty Stewart rests on the beach beside one of Alex and Vicki Mills's private Miramichi River pools. PHILIP LEE

The private ownership of rivers has its roots in Europe, where water rights go back hundreds of years. When settlers first came to New Brunswick, they were granted exclusive fishing rights as a way of ensuring a food supply for their families. In 1894, the New Brunswick government began to reserve some fishing rights for the Crown along the shore of major rivers. In 1925, the province stopped granting fishing rights altogether, although some exclusive fishing grants slipped through after that date.

Who owns what along salmon rivers — what water is private and what is public — is an ongoing debate in New Brunswick, with as many as 50 requests for searches coming each year to the Crown Lands branch of the Department of Natural Resources. River ownership in New Brunswick breaks down this way: 44.5 per cent (1,078 kilometres) is privately owned; 33.5 per cent (816 kilometres) is Crown-owned open water with unrestricted public fishing; 8.1 per cent (197 kilometres) is Crown Reserve water offered to New Brunswick anglers only through a lottery conducted each winter; 7.2 per cent (174 kilometres) is Crown-leased; and 6.2 per cent (146 kilometres) is Crown-closed water (sometimes open for hook-and-release fishing only) to provide a sanctuary for fish. To apportion Crown-leased water, the Department of Natural Resources holds a public auction every 10 years; the winners of the leases have exclusive fishing rights on those sections of river. These leases include the riverbed and one rod (16 feet) of upland. Strict conditions apply: the leaseholders must hire people to clean up along their stretch of riverbank and employ wardens to guard 24 hours a day against poaching and illegal fishing. Nineteen 10-year leases were auctioned off in 1989.

The headwaters and important spawning tributaries on the Miramichi system are privately owned and protected by two timber companies — J.D. Irving Ltd. and Avenor Inc., which own the river from 10 kilometres above Boiestown up to Juniper (about 80 kilometres of river). The upper half is owned by J.D. Irving Ltd.; the lower half, once owned by Canadian Pacific and International Paper, is now owned by Avenor. Avenor runs a salmon fishing camp at the mouth of Rocky Brook, an important spawning tributary. Access to the area and fishing are strictly controlled. The Irvings control access to Clearwater Brook, a pristine spawning tributary, and the Alder Grounds, where the North Branch of the Main Southwest Miramichi breaks into a maze of small streams.

John Gilbert, J.D. Irving Ltd.'s manager of fish and wildlife habitat, says gates across logging roads that restrict access to the headwaters of the salmon rivers have preserved important spawning sanctuaries. For example, once salmon make it to Boston Brook on the Little Main Restigouche River, a tributary of the Restigouche River where the Irvings have a camp and riparian rights, they are lightly fished; fewer than a dozen people will fish there during a season. "Once the salmon hit there, they're safe," Gilbert says. "We know from experience that easily accessible streams and lakes, where fishing is uncontrolled, are soon fished out. If you want to kill a lake, put a road by it."

J.D. Irving Ltd. also owns and leases other Miramichi and Restigouche waters, including the Downs Gulch stretch on the Restigouche (a Crown lease) and the famous Big Hole Brook Pool on the Main Southwest Miramichi. The Big Hole deep-water holding pool is an important fish sanctuary where thousands of salmon rest at the mouth of a cold-water stream during the dog days of summer. Irving wardens, based at a neat little camp on a high bank overlooking the pool, guard the salmon day and night.

Atlantic salmon angling in New Brunswick is a big business that brings in more than $50-million a year during the fishing season. Jack Fenety says salmon water real estate in New Brunswick is worth at least $500-million — as long as salmon continue to return to the pools. Big Hole Brook, which is one of the best salmon pools in the world, is worth more than $1-million.

New Brunswick's system of riparian rights can be traced back to Scotland, which has a long tradition of private ownership of rivers. Robert Williamson believes that property rights can be a powerful conservation tool. "The salmon is a migratory fish, and its movements are predictable. Because it returns to rivers, it comes to places where it's easily accessible and as a consequence it can be relatively easily caught. The right to fish for salmon was particularly valuable. If access to a renewable resource is private property, then the proper, long-term management of that renewable resource is in the direct and immediate interests of those who have the right of access to it. Where there is a history and a natural development of private ownership of fishing rights, then there could be advantages in ensuring that you don't lose it, or in making it work as well as you can."

One of the earliest pollution laws in Scotland was part of a salmon act in the 1850s. It was a private act promoted by the salmon fishery owners, who said noxious elements should not be allowed to flow into any salmon river. "That was promoted by the salmon fishery owners in order to protect their asset and they persuaded Parliament that it should be passed and it was passed," Williamson says.

Elizabeth Brubaker, the executive director of the advocacy group Environment Probe, says in her book *Property Rights in the Defence of Nature*, "In a system of secure property rights, ownership also promotes stewardship. Individual, corporate, or community owners have incentives to maximize their resources' value, taking into account not just their current worth but their future value as well. The guardian of Canada's fisheries — the federal government — has likewise overseen that resource's destruction. Communities, firms and individuals who relied on but did not own these fisheries have been powerless to save them."

*

Riparian owners in New Brunswick have taken action to protect their waters when the federal government seemed to be overseeing the extinction of the Atlantic salmon. In the early 1980s, when the ocean commercial salmon fishery was threatening to wipe out salmon runs to New Brunswick rivers, riparian owners like Alex and Vicki Mills organized themselves to defend their waters. On Aug. 3, 1983, 300 riparian owners and citizens of the Miramichi met at the Doaktown High School.

"It was wild," Vicki Mills recalls. "We had hundreds of people. The place was packed. All of them were just beside themselves. We were picking up the paper and reading articles about the commercial fishermen bragging about taking 10 times their limit and getting away with it. [The Department of Fisheries and

Alex and Vicki Mills outside Old River Lodge, their fishing camp on the Miramichi River below Doaktown, New Brunswick. THOMAS R. PERO

Oceans] had no recognition of the sport fishery or the importance of the economics."

The economics were simple — the commercial fishermen, who had no ownership stake in the resource other than a fishing licence, were killing 88 per cent of the fish and generating only 15 per cent of the Atlantic salmon revenue. Commercial salmon netsmen had seen their quotas cut back in 1972 (Ottawa paid them more than $13-million in compensation during this period), but they were still allowed to keep "incidentally caught" salmon, fish that they happened to catch while fishing for other species, and fishermen continued to make good money off salmon landings. In 1981, then-federal fisheries minister Romeo LeBlanc (now Governor-General) lifted the salmon ban, allowing netsmen to catch 100 salmon each. This fishery guaranteed widespread poaching, because the low number of salmon per fisherman didn't allow a netsman to pay his expenses.

In the summer of 1983, Ted Williams, the baseball legend, angler and fierce Atlantic salmon conservationist, told journalist Thomas R. Pero (who was writing about the Miramichi for *Trout* magazine) that Romeo LeBlanc "did more to kill this beautiful fish than anyone I know." Williams, who had been a fixture since 1961 at his fishing camp near the town of Blackville on the Miramichi, suggested that LeBlanc's picture be added to the display of the salmon's natural predators — the otter, merganser and osprey — at the Atlantic Salmon Museum in Doaktown. "I told them I'd hand them a thousand bucks on the spot if they'd put Romeo LeBlanc's picture right up there in the middle," Williams said.

In 1983, commercial fishermen were given a quota of 114 fish each. Some fishermen publicly boasted that they were averaging 350 to 400 salmon each. That's when the riparian owners and people who had stakes in the river took action. Vicki and Alex Mills, Vincent Swazey, Herb Wade of Wade's Fishing Lodge and Keith Wilson of Wilson's Sporting Camps went to Ottawa in the winter of 1984 to present their case. Ottawa listened to them and to the demands for action from other riparian owners throughout New Brunswick. In the 1984 season, commercial netsmen in New Brunswick saw their season reduced and the incidental catch rules eliminated. Anglers were limited to killing only grilse; large salmon had to be released alive, a regulation that is still in force and widely accepted today.

In April 1985, then-fisheries minister John Fraser banned commercial netting in the Maritimes and cut back Newfoundland netting further. Ottawa began to spend millions of dollars buying out the licences of the public ocean fishery.

Fraser, a salmon angler who often fishes the Miramichi, is a hero among salmon conservationists. In 1985, he said: "I'm concerned first and foremost with the Atlantic salmon resource." The owners of salmon waters were behind him 100 per cent.

Vince Swazey and baseball legend Ted Williams prepare for a morning on the Miramichi River. THOMAS R. PERO

*

On a summer day along any of New Brunswick's salmon rivers, when you're standing along private or leased water, you may hear a familiar shout from a passing canoe — Yankee go home. "People say the Miramichi is owned by rich Americans. Well, that is total hogwash," says Alex Mills. "There is a very substantial part of the Miramichi that is owned in 1995 by the same families who owned it in 1895. A lot of owners on the Miramichi are local residents." Some of the best waters, the most productive waters, have indeed been sold to anglers and corporations in the United States, but Alex Mills notes that it was New Brunswickers who sold their fishing rights. "I never saw any local here get skinned on a piece of water. I think private ownership is what has saved the salmon, if for no other reason than that you have a lot of guys with a lot of money who can own private water, and they have been prepared to put the big bucks up to fight for salmon conservation. As for managing, private ownership controls access, but I think the real plus is that the owners of the private waters are the guys who have put up the huge bucks, tens of millions of dollars over the years, to fight for salmon conservation."

Mills notes that the Atlantic Salmon Federation, which for decades has led the conservation movement and the push for more salmon research, wouldn't exist without donations from private river owners. The Miramichi Salmon Association is Canadian-controlled but heavily supported by donations from south of the border.

"Nobody made those Canadians sell those waters to the Americans," Fenety says. "They did it voluntarily to attain a capital gain. They sold the old homestead that had riparian rights. The Americans come here because they've destroyed their own salmon rivers and they're happy to share what we have. More than 70 per cent of all the money that goes into salmon conservation is put up by Americans. It's a shameful thing, but that's true. If that was an open piece of water, you might have 85 people on one shore and 90 on the other and they'd take every last fish that they could take. And this is just human nature. The basic problem is that we will never have enough water, enough rivers, enough pools, enough salmon to satisfy the public demand. Someone says I should be able to come and fish your water. I say that's great. You own 40 acres of land and I should be able to go out there on your land and cut the wood to build my house. What's the difference? I own this and you own that. You can't have it both ways. Private ownership means private ownership."

Some New Brunswick residents have resented the fact that there are private pools on their rivers. They think that for the cost of a salmon licence ($17.12 in 1996) they should be able to fish anywhere they want all season long. Swazey says, "Most of our local people feel that they should be able to fish for nothing. It isn't the way of things. It costs a lot to have private pools for your guests to fish in, both in taxes and leases. You can't go golfing for free. I don't know of anything that you can go and enjoy that isn't going to cost you something. Fishing shouldn't be any different from anything else. It's entertainment. There isn't room enough for everybody who wants to go salmon fishing. And I don't know where you draw the dividing line unless it's the people who can afford to pay their way. There's got to be somewhere to draw that line."

✻

In the United States, where private ownership of rivers is rare, the belief in free public fishing is strong and fishing pressure is high, fish populations are being preserved by catch-and-release restrictions. Howell Raines, the editorial page editor of the *New York Times*, writes in his fine book, *Fly Fishing Through the Midlife Crisis*, that without mass acceptance of catch-and-release fishing, the best trout and bass fishing in the U.S. would have been wiped out by the year 2000.

> I speak as a person who has seen the great and seemingly inexhaustible snapper, grouper, sea trout and mackerel fishery of the northern Gulf of Mexico decimated in his own lifetime. A charter captain along the "Redneck Riviera" summed it up perfectly when asked what happened to the red snapper that used to inhabit the reefs off Panama City, Destin and Pensacola. "They all went to Birmingham and Atlanta in ice chests." To a person who does not follow fishing, it seems impossible that our vast lakes, rivers, streams, bays, estuaries and bordering oceans could be picked clean. But remember that in the seemingly in-

> exhaustible plains of Montana, a handful of hunters using primitive weapons reduced the buffalo herd from 13 million to two hundred in only five years between 1876 and 1881.

Lee Wulff, a pioneer of catch-and-release fishing, maintained that uncontrolled public fishing would destroy the Atlantic salmon. He wrote in *The Atlantic Salmon*: "There is considerable conflict of ideas as to whether Atlantic salmon fishing waters should be open to the public, owned outright by individuals or controlled in some manner between these two extremes. Unlimited public fishing for salmon, where the licence fee is low and where salmon waters are readily accessible to anglers, soon results in very poor salmon fishing, or none at all."

The Commons

Last night I dreamed that I was sailing
Down on the Sea of Galilee.
We cast our nets upon the water
Jesus pulled them in with me.
— *Lennie Gallant*

In his song "Peter's Dream," Maritime songwriter Lennie Gallant tells the story of Peter, a small-boat fisherman who gets up before the day breaks and walks down to the ocean shore. Peter watches the sun rise and then turns and walks back towards his home. His fishing boat, *The Maiden*, doesn't sail anymore. He dreams of fishing with Jesus on the Sea of Galilee, where he can cast his empty net on the other side of the boat and find it full of fish. In Peter's world, the fish are gone.

Small-boat fishermen in Newfoundland are in the unthinkable position of being arrested for trying to catch fish in the harbours to feed their families. How could this have happened? Former federal fisheries minister John Crosbie, an intelligent, combative Newfoundlander, knows the history of the East Coast fishery as well as anyone in Canada. He described the central problem facing the ocean fisheries when he was part of a panel organized by the *Telegraph Journal*: "It's the selfishness of those people who are out there fishing. They're all selfish because it's first-come, first-served. Whoever gets their hands on the most fish the quickest gets the greatest benefit. That's what makes the fishery so difficult to regulate.

And our own fishermen are not one damn bit better than the fishermen of any other country I've ever come across. They're all into the same game. The central issue in the fishery which has never been resolved, the greatest problem, is that the fish are owned by no one. It's the problem of the commons: Nobody owns any of the fish until they're taken into their possession, and that leads to many of the problems that we've had."

The fishery on the commons, our publicly owned, government-managed ocean, has produced too many fishermen, too many fish plants and too many fish-plant workers. "At least half of them will never work in a fish-processing plant again, but they're not being told who that is," Crosbie said. "We've got too many fishermen, and they're in the same position. So this is a tremendous problem, a political and economic and social problem."

The destruction of the northern cod stock — once the greatest fish stock in the world, a stock that numbered in the hundreds of millions — is Canada's environmental crime of the century. The northern cod was fished out by Canadian and foreign fishermen, much of the damage being done by high-tech trawlers that aggressively fished on the high seas, slaughtered fish on their spawning grounds and tore up the ocean floor.

"Times had certainly changed since John Cabot's voyage of discovery in the 15th century, when cod were taken in profusion simply by lowering a weighted basket over the side of his wooden ship and pulling it up on a rope," writes Michael Harris in *Rare Ambition*, his history of the Crosbie family in Newfoundland. "Now the silver mine of the North Atlantic was played out and, while no one was prepared to own up to their own role in the pillage, all were quick to wag an accusing finger at a host of other villains. The Canadians blamed the foreigners; the foreigners blamed the Canadians; the domestic inshore and offshore fleets blamed each other and everyone blamed the federal government — Atlantic Canada's old, sad song."

Overfishing took its toll and harsh environmental conditions battered the stock at the same time. Although scientists warned that fishing had to be drastically reduced, Ottawa kept the total allowable catch too high, gambling on the future of the resource.

"While the fish disaster brewed, keeping people employed dominated the government's concerns," writes Elizabeth Brubaker. "Managers based catch limits not just on what they thought the fishery could bear, but also on the economic needs of fishing communities. Essentially the government was paying people to destroy a valuable resource by enabling them to stay in an industry that couldn't support them."

When Ottawa finally declared a moratorium on cod and capelin fishing in

1992, the fishery had reached such a low point that some scientists were not sure that it would ever recover. It was a social and economic disaster — 30,000 fishermen and plant workers were out of work. The mighty northern cod stock had been reduced to just three per cent of its spawning capacity.

*

In the last century, the world ocean catch of fish has grown from five million tonnes to more than 80 million tonnes a year. The United Nations Food and Agriculture Organization says that all 17 of the world's major fishing areas have either reached or exceeded their natural limits and that nine of these areas are in serious trouble.

Deborah Cramer, a Massachusetts author who is writing a book about the Atlantic Ocean, describes the sorry state of the ocean fisheries in the June, 1995, issue of the *Atlantic Monthly*. She writes that Georges Bank off New England, once one of the world's most prolific producers of cod, haddock and yellowtail flounder, is nearly empty of fish. The United States National Marine Fisheries Service has ordered temporary fishing bans while it looks for a longer-term solution. Cramer writes: "Fish are like a savings account or an investment. With enough capital it's possible to live off the interest. Fishermen working New England's waters have not only spent the interest but also squandered the capital, taking an average of 60 per cent of the entire population of cod, haddock, and yellowtail each year."

In the North Sea, an ocean that provides fish for 65 million people in seven European countries, fishermen compete for haddock, cod and sole. "Fishing pressure in the North Sea is so intense that not even one per cent of the one-year-old cod remain in the sea long enough to spawn," Cramer writes. "Since catches of cod now consist primarily of immature fish, the prospects for the survival of the population are not good."

North Atlantic fishermen are now turning to the waters off Africa and to the Indian Ocean — half the fish caught off the coast of Africa are now taken by non-African fleets. And fish-killing technology continues to improve. The mouth of the latest type of trawling net is large enough to catch 12 Boeing 747 jets in two piles of six.

*

Twenty years before Ottawa closed Newfoundland's Atlantic salmon ocean fishery, Jack Fenety was campaigning against the philosophy of fishing on the commons. "[The fishermen] were people who didn't have to buy land, who didn't plant, didn't harrow, did nothing but harvest," he says. "Their sole aim and objective was to harvest. I used to say this was a dangerous thing and ultimately only destruction would result." He recalls speaking in Newfoundland to a group of commercial netsmen, asking them what they planned to do when the salmon were all gone. They answered: "By the Lard Jesus, sir, we'll fish for something else." Now they don't even have cod to fish.

The Atlantic salmon, unlike the cod, was not completely wiped out in the ocean — because river owners demanded that their assets be protected. They had a long-term investment in their waters and they were not willing simply to take what they could get out of the resource and then move on to something else.

"Fortunately, the government of Canada took action before the salmon went over the stock precipice that the cod did," says Dr. John Anderson, the Atlantic Salmon Federation's vice-president of operations. "They took action by introducing harvest regulations for anglers, but most importantly they took the bold step of taking the commercial nets out of the water. That required a fair amount of political courage and a lot of money to compensate the commercial fishermen. Had that not happened, then probably the Atlantic salmon would be like the cod, something we could talk about in historical terms, and that's about it."

Critics of the concept of protecting our resources with secure property rights argue that such a system ensures wealth for a powerful few. However, it could also be argued that the commons ocean resource never distributed wealth fairly. The Newfoundland fishery historically produced a wealthy merchant class and an underclass of poor fishermen. Harris writes in his Crosbie family history: "The result was an isolated and dependent outport population that in good times was controlled by the merchants of Water Street and in bad times relied on government to stave off disaster — a never too distant possibility in a rural economy based on the vagaries of the inshore fishery and the thin harvests of subsistence agriculture."

Although government scientists and managers made mistakes in the way they managed the northern cod, to say they were utterly incompetent is wrong. On the other hand, to say we simply need better government management is naive. The northern cod and other fish stocks were destroyed despite a massive scientific and management bureaucracy. The fishery has been the subject of two major commissions of inquiry in the past three decades, and in 1995, there was one full-time Department of Fisheries and Oceans employee for every six full-time fishermen in the Maritimes. The roots of this environmental tragedy run much deeper. If we look back at the course the Newfoundland fishery was taking before Ottawa took firm control of the wheel and drove it to the bottom of the ocean, we find the beginnings of the tragedy, and perhaps the way to a sustainable fishery in the future.

❋

In his study of the collapse of the Newfoundland inshore cod fishery, David Ralph Matthews, the associate dean of Graduate Studies at McMaster University in Hamilton, challenges the notion that the Newfoundland fishery was destined to become another tragedy of the commons. When researching his book *Controlling Common Property*, Matthews discovered that small Newfoundland communities traditionally regulated their fishery through a kind of property rights system, carefully guarding their fishing grounds against intruders from other communities.

Within the community fishing grounds, individual fishermen would have their own berths for their trap nets, to which they would return year after year. Communities would set restrictions on what type of fishing gear could be used in certain berths, trying to balance their fishing space with the number of fishermen. Matthews argues that Ottawa's decision to manage this traditional fishery like a commons fishery led to ruin.

According to Matthews, Ottawa decided that the East Coast fishery was an unstable common resource and chose to try to limit access to it through licensing. This policy simply meant more fish were being caught by fewer (and only slightly fewer) fishermen. At the same time, in the 1980s, the government of Newfoundland under Brian Peckford was vehemently opposing any restrictions on the number of fishermen: "The provincial government has taken the view that the right to fish is a local birthright." Peckford was willing to limit types of fishing gear and the number of large boats, but not the number of fishermen.

Matthews advocates a return to a form of community ownership for the commons fishing grounds. "[O]ne solution to the problem of fisheries management at the community level would involve the state formally allocating to each community the property rights in its local community fishing grounds that the members of that work community already believe they possess." These property rights must be accompanied by management powers. "If property rights in the inshore fishery are to be formally lodged within the work communities themselves, those who fish from these communities must be given real power to decide on allowable catch allocations, appropriate use of gear, licensing regulations and a variety of other dimensions of fishing practice. In other words they must have more than a simply advisory role."

Brubaker advocates the creation of long-term, transferable property rights in Canada's fisheries, a system of individual quotas that guarantee fishermen a share of the total catch. "Transferable quotas provide a number of incentives to use resources efficiently. Confident that their rights to fish are secure, fishermen need not waste money building bigger boats and equipping them with more advanced gear in a race to catch fish. These quotas also promote conservation. With valuable assets tied up in their property rights to a percentage of the catch, fishermen have an economic interest in conserving fish stocks and in keeping out interlopers; it is they who will capture the benefits of such activities."

The National Marine Fisheries Service (NMFS) in the United States launched a test case for Individual Transferable Quotas (ITQs) in 1990 to try to stop overfishing of the ocean quahog (a type of clam used in chowder) and surf clam populations. Writing about the new system in *Harper's* magazine, Washington journalist Bill Gifford says the move to individual quotas "represents a drastic departure from traditional fisheries management, which relies upon restricting gear and limiting seasons, tactics that have proven largely ineffectual. By the time NMFS imposed ITQs on quahogs and surf clams in 1990, overfishing had forced the feds to restrict harvesting to six hours every two weeks. Fishermen responded by buying bigger boats and sailing in gales to best their competition, causing debt, deaths, and the fisheries' continuing decline. ITQs brought this 'race for fish' to an end, but with it centuries of open-access fishing. Before, anyone could dredge for clams; now only quota owners (or renters) may do so." Gifford, who laments the shift of the fisheries management debate "from preservation to profit," notes that the National Marine Fisheries Service is bringing individual quotas to the Alaskan halibut and red snapper fisheries. He predicts that the fight over who receives the quotas in America's largest fishery, pollack, will be a "bloodbath."

Brubaker points out that technology may make it easier to assign private fishing rights in the ocean; NASA has already been experimenting with technology that can monitor the oceans with satellites, identifying boats by the fingerprints they leave behind in their exhaust. "Broader property rights in fisheries, as in other resources, would be a two-edged sword," she writes. "Inevitably, some owners would either through error or ignorance deplete their stocks. As long as their activities did not violate others' property rights, the law would permit them. But with the removal of governments' perverse incentives to harvest uneconomically, the destruction of private property would occur less frequently. That which did occur would be modest compared to the wholesale environmental devastation wrought directly or abetted by governments."

*

The French Revolution provides a final analogy. In France, before the Revolution, the right to fish in rivers was a royal right sometimes extended to clerics and members of the nobility. The public wasn't allowed to fish for salmon. Although the system was manifestly unfair, there were fish in the rivers. Anthony Netboy writes:

> Landowners seem to have been usually solicitous of migratory fish and took pains to keep the waterways unobstructed and assure passage at difficult places. Kings of the Middle Ages, beginning in the reign of Charlemagne, passed statutes regulating fishing with an eye to preserving the stocks. Until the Revolution of 1789, at least, the freshwater fishes were generally in ample supply, subject to normal fluctuations of abundance. Fishing seasons were reasonably observed, and waterways were usually well tended. Dissipation of the inland fisheries began with the French Revolution. Decrees of the Legislature and Convention of 1792 and 1793 abolished the exclusive rights of the Crown, Church and nobility to hunt and fish on the lands of the kingdom. The Republic proclaimed free fishing for all citizens on all rivers. Hitherto barred from sport by high walls and harsh trespassing laws, the populace began to pillage the resources. "What belongs to everybody belongs to nobody," was a common saying; hence, citizens helped themselves to all the fish and game they could obtain. Fishing rights were later restored to the state and riparian owners regained their rights on some waters. But the damage was done.

Netboy asked Robert Bachelier, the official in charge of France's inland fisheries, to explain the destruction of France's Atlantic salmon runs. Bachelier replied: "Before the Revolution there was respect for private property in France. During the Revolutionary era, people went wild, hunting and fishing without restraint. That spirit never really died." Netboy quotes authors L. de Boisset and Richard Vibert, who in 1944 lamented the loss of river fish stocks in France in their book *La Peche Fluviale en France*: "Woe to the guardians of the public wealth who, through ignorance or the crime of carelessness, have permitted the property of which they were the keepers to slip through their hands, since they can be sure that God on the Day of Judgement will make them pass on the left side."

Riparian Rights

Eighty years ago, the members of the Nepisiguit Real Estate and Fishing Company had riparian rights on the Nepisiguit River in northern New Brunswick. The Nepisiguit is a famous salmon stream where large fish fill rocky pools along 28 kilometres of water that runs from a 30-metre waterfall into the sea. In the summer of 1912, the Canadian Iron Corporation, mining upstream on one of the Nepisiguit's tributaries, polluted the river by allowing water to leak into the stream while it was washing and grinding ore. When pollution spread downstream into the salmon pools, the riparian owners went to court, arguing that the pollution was ruining Atlantic salmon spawning grounds and making fishing impossible. The corporation fought the charges.

The judge, who visited the site on the river and saw the discoloured water, ruled in favour of the fishermen. In his judgement, he said, the fishing club, as a riparian owner, had "the right to the full flow of water in its natural state, without any diminution or pollution." The judge gave the corporation four months to clean up its operation.

Riparian owners today, like the Nepisiguit anglers of 1912, have powerful legal weapons to defend the quality of water in their rivers and to protect Atlantic salmon. The water rights bible for Atlantic Canada — *Water Law in Canada: The Atlantic Provinces* — was written by a New Brunswicker, Gerard V. La Forest, now a justice of the Supreme Court of Canada. La Forest lays out in detail the impressive common-law rights of a riparian owner, a person who owns land that borders on a body of water, whether that be a river, stream, lake or the sea.

The right of access is the riparian owner's most basic right, and owners can bring a court action against anyone who blocks access to the water. "No one, not even the Crown, can erect any structure on the shore or otherwise permanently obstruct a riparian owner's right of access," La Forest writes. "The riparian owner's right of access exists in a direct line from every point along the whole frontage of his land on the water."

Riparian owners can drain their lands into the stream, but this has to be reasonable drainage and can't pollute the river. "The law regarding pollution is clear," La Forest writes. "If a landowner pollutes the water, he is liable for damages and he may be enjoined from doing so by injunction."

The riparian owner has the right to clean water, "the flow of water in its natural state — unpolluted," and the right to receive water flowing in its natural course. The upper owners aren't allowed to take water from the stream permanently, and the flow to property downstream can't be changed. A riparian owner may change the course of the stream on his own land as long as it flows back into

its normal channel. "There is nothing to prevent a landowner from straightening, cleaning out or deepening the channel on his land. Similarly, the diversion of the waters of a stream to a flume or mill-race to operate a mill is permissible. So too is irrigation. But if the diversion affects the flow to the detriment of a riparian owner downstream, the upstream owner will be liable." The water must return to the stream "substantially undiminished in quantity and quality." The riparian owner also has the right to have water leave his land without obstruction — for example, a river can't be backed up on a riparian owner's land by a dam or bridge downstream.

Anglers fish one of the Nepisiguit River's rocky salmon pools. ATLANTIC SALMON FEDERATION

The riparian owner doesn't need to suffer damages to obtain an injunction to stop pollution, and the social or economic importance of the polluter will not affect an injunction. "Pollution of streams frequently arises in connection with drainage and sewage systems of municipalities," La Forest writes. "There is no question that municipalities, like other landowners, have the right to drain their lands, but this gives them no right to pollute streams."

Not all riparian owners have private fishing rights. In tidal waters, the public has a right to fish. Above tidal waters, the owner of the streambed has the exclusive right to the fishery, and this right comes with a set of responsibilities. "A person who owns a fishery must himself exercise his rights reasonably. Though he may, subject to the fishing laws, catch as many fish as he can, he must not in exercising his rights do anything to injure the rights of persons above or below him." Canadian courts have said that the owner of the land through which a stream flows owns the bed, unless it has been reserved by the Crown or sold. If a person owns one bank, that person owns the stream bed to the centre thread of the stream.

Some riparian rights depend on principles of Roman law; others evolved under English common law. Riparian rights developed rapidly in the British courts as river owners fought the onslaught of water pollution during the Industrial Revolution and the introduction of piped municipal sewage systems in the 1800s. Brubaker writes: "People fought back, taking their cases to the courts, which gradually honed the law into a finely-tuned system of rights and responsibilities. By the 1850s, when the courts were awash in water pollution cases, the riparian doctrine had emerged as a powerful weapon against the era's severe environmental challenges. Waterfront property owners, businesses, and others dependent on lakes and rivers fought industrial and municipal water pollution with riparian law throughout much of the English-speaking world during the following century, and in some places continue to do so."

One famous riparian court case in Great Britian involved a fight over Scotch whisky and showed just how potent riparian rights are in protecting the natural qualities of fresh water. (A series of riparian cases, including this one, are summarized by Brubaker in *Property Rights in the Defense of Nature*.)

The Bankier Distillery Company in Scotland was making whisky using pure, soft water from a stream called Doup Burn. In the 1890s, John Young and Company started pumping water out of its coal mines into Doup Burn. The water was still drinkable, but it had become too hard for the whisky-making process. Bankier Distillery went to court arguing that its riparian rights had been violated. In 1893, the House of Lords upheld a lower court's decision to stop the mine water from going into the burn, despite the fact that the water was still fit to drink. Lord Shand said: "The lower owner is entitled to have the water transmitted to him with its natural qualities unimpaired." Lord Macnaghten noted: "It is not permissible in such a case for a man to use his own property so as to injure the property of his neighbour."

In 1948, John F. Eastwood, a barrister and angler in London, England, founded the Angler's Co-operative Association. Since then the association has gone to court on behalf of riparian owners and fought and won hundreds of cases. "They were determined to fight polluters and pollution by asserting common law rights," writes Anthony Netboy. "Anti-pollution laws seemed to have been of no practical use to victims of pollution. Through its vigorous pursuit of the rights of owners, the association has forced the courts to rule that anyone — industry, local government, farmers — who uses the rivers as the cheapest way of getting rid of wastes is breaking the law."

In the early 1950s, the Pride of Derby and Derbyshire Angling Ltd. and the Earl of Harrington went to court to save their salmon river, the Derwent, in north central England. They took three upstream polluters to court. The Borough of Derby was dumping sewage into the river; British Celanese Ltd. was polluting and warming the water; and the British Electricity Authority was increasing the water's temperature. The angling club and the Earl won a court injunction ordering the three polluters to stop polluting, though it allowed them two years to clean up their act. The Borough of Derby argued that the population had outgrown the sewage system. Lord Justice Denning replied that municipal authorities "know (or ought to know) that the increase in building will cause the existing sewers to overflow, and yet they allow it to go on without enlarging the capacity of the sewerage system. By so doing, they themselves are helping to fill the system beyond its capacity, and are guilty of nuisance." Nor would the court let the Borough of Derby substitute damages for the order to rebuild its sewage system: "Anyone who creates an actionable nuisance is a wrongdoer, and the court will restrain him from persisting in his activities."

A Canadian case emphasized that unless parliament or a legislature decides otherwise, public works must not interfere with the private rights of individual property owners. In the 1920s, an Edmonton family took the city to court for fouling a stream that flowed through the family's property. The city had been flushing horse droppings and street filth into the stream. The case ended up before the Supreme Court of Canada, where the judges decided the family's riparian rights had been violated. The city claimed it had the statutory authority to pollute; the court did not agree. Statutory powers do not authorize a private nuisance "unless indeed the statute expressly so states." The high court granted damages and an injunction against further pollution.

Brubaker points out that riparian law, like other common-law doctrines, survives only when politicians haven't taken these rights away through statutes and regulations. "Ownership empowers people; legally enforceable property rights invest owners with the authority — and the tools — to protect the fisheries and the waters they control."

By contrast, the users of a public, commonly held property are powerless when a resource is threatened. In a notorious Newfoundland case in 1970, fishermen in

Placentia Bay launched a court action to try to stop a phosphorus plant from dumping poisons into the sea that were killing the fish. Brubaker writes: "Although the court acknowledged that the pollution, by harming the fish, may have jeopardized the fishermen's livelihoods, it decided that they lacked the standing to sue: First they had no property interest in the fish, and second, their damages did not differ in both kind and degree from those of the general public, all of whom had the right to fish in public waters and all of whose rights had been violated to the same degree."

In 1991, when the Canadian government proposed entrenching property rights in the constitution, environmentalists loudly objected, saying this would give people the right to pollute. However, if property owners could fight cases to the Supreme Court, governments and potential polluters would think twice before they compromised the environment. Brubaker writes: "The more precisely property rights can be defined, and the greater the extent to which they are assigned to specific groups and individuals, the better the environment will fare. If given the proper tools, owners will protect their resources. It is long past time for responsibility for natural resources to be shifted away from governments and back to the individuals and communities that have strong interests in their preservation."

*

Cultural ecologist E.N. Anderson, in his fine book *Ecologies of the Heart: Emotion, Belief and the Environment*, argues that solutions to our environmental problems will come from a mixture of reason and passion. It is "genuinely immoral" to destroy a species or an ecosystem, something that we cannot recycle or re-create, he writes. "Any management strategy that does not take human feeling into account simply will not work." To save the environment, people must develop "some form of environmental morality," and the political process "must accommodate environmental goals."

Anderson believes that the revival of local management is a workable option for protecting the environment, whereas highly centralized, authoritarian governments "are always a catastrophe." "The farther control gets from the local people who actually benefit or lose by resource management, and the more control is vested in people remote from the scene, the worse the management." Big governments, remote governments, "tend to get out of touch with local conditions and with people on the ground."

People need a legal recourse; they must be able to sue those who damage their resource holdings. "The coercive role of government is limited. Neither is coercive government desirable. Thus the current environmentalist tendency to rely on big government, and heavy regulation, is dangerously misguided. We cannot and should not put a soldier behind every tree and every bird. Even if we did, it wouldn't work; if society reached the point at which such a course was necessary, nothing could save the environment. Thus, solutions must be found in motivating individuals, especially those individuals who actually use the resources, to conserve." Riparian rights give river owners the legal clout to protect their holdings, and they also demand that the owners themselves be responsible. A regime of riparian rights and local management on our salmon rivers — a system that combines reason and passion — would move us a step towards Anderson's enlightened vision.

CHAPTER THREE

In Our Hands

Senseless slaughter

It is a curious situation that the sea, from which life first arose, should now be threatened by the activities of one form of that life. But the sea, though changed in a sinister way, will continue to exist; the threat is rather to life itself.

— Rachel Carson

When tiny silver capelin, millions strong, rolled in like waves to spawn along the gravel beaches that dot Newfoundland's rugged coastline, their arrival was a public event. Families headed for the shore. Children waded into the waves and filled buckets to the brim with sparkling fish. Their parents scooped up bags of capelin to take home — some fish to eat and others to fertilize their gardens. When the capelin arrived, inshore fishermen knew the northern cod would be following fast behind, returning from their offshore spawning grounds to feed on the shoals of small fish.

In *The Foxes of Beachy Cove*, Newfoundland author Harold Horwood describes the magic of the capelin as he watched several million small fish "struggling in the inch-deep water" on the edge of the shore at Beachy Cove. "Each evening after that we waited on the beach, cast nets ready, for the coming of the tide. Then, as the sun set and the sea became milky and obscure, someone would proclaim excitedly, 'Look there — right off the mouth of the brook!' And there they would be, packed together in unbelievable masses, turning and flipping and struggling in the gentle surf, then as the surge receded, somersaulting breathless over the sand toward the retreating line of the sea."

In the 1980s, the capelin were rounded up in huge nets to fetch top dollar on the Japanese market. By 1991, they had disappeared. "Capelin do not come in to the beaches to spawn in any numbers any more," writes Ches Loughlin, editor of *Spawner*, the annual magazine of the Salmon Preservation Association for the Waters of Newfoundland. "Most of us have not been able to find any for three or four years or longer. Where once they rolled to spawn in millions, now they can hardly be seen at all. The capelin that do roll are too small to eat. Any biologist worth his salt will tell you that when the spawning stock shrinks to the size that only small, immature fish are available to spawn, the body of the fish is under great stress." Any biologist worth his salt will tell you there's something seriously wrong in the ocean.

Marine biologists call the ocean the black box. The Atlantic Salmon Federation's Dr. John Anderson says: "It's black in the sense that you can't see very much and also black because we don't know what the heck is happening out there. We are getting perilously close to there not even being enough salmon, if there's no harvest whatever, to look after seeding the rivers. For the last six years there has been a decline that cannot be attributed to fishing mortality. It's due to natural mortality. It's possible that man has interfered with the natural order of things and therefore the whole food web is not what nature had in mind at all. The feeling I have is that if we can really nail down what's happening to Atlantic salmon, why that survival is not the way it should be, it probably would be applicable to other species. Whatever we find out about Atlantic salmon would help shed some light on some dark corners in this thing out there we call the black box."

Tracking fish in the North Atlantic is far more difficult than counting fish that return to a river. Smolt leave their rivers in predictable ways. Once they leave the rivers, how do you follow the fish? "Imagine trying to assess the stock of forests in Canada but you can't see the trees because you're up in an airplane and all you see are clouds," Anderson says.

In *Wild Steelhead and Atlantic Salmon* magazine, two United States biologists addressed the question of what is happening to the salmon at sea. Allen E. Peterson Jr., the science and research director for the National Marine Fisheries Service at Woods Hole, Massachusetts, and Dr. Kevin D. Friedland, a specialist in salmon migration who also works at Woods Hole, concluded that the salmon are dying during their first year at sea. "Consider this: For every 10,000 fish stocked in the Penobscot River in Maine, approximately 110 would have been caught at Greenland or Newfoundland. One hundred or so might come back to the river. After accounting for natural mortality in older fish, we are left with 9,780 salmon that perished before the fisheries began. What happened to them?"

The two biologists argue that the decline in salmon at sea is due to cyclical environmental changes that are reducing the feeding area for the salmon. They say that when ocean conditions improve, Atlantic salmon numbers will increase. "Salmon runs to our rivers have been good and bad over recorded history," they write. "In pristine environments, before disturbance and destruction by modern civilization, these fluctuations were barely noticeable because Mother Nature has a propensity to compensate for natural losses by creating excess production." But today there are many unnatural forces at play, and we face daunting environmental and management problems both in the ocean and in our rivers.

❋

The ocean's food chain and ecosystem may be in collapse. Ches Loughlin points out that capelin are a major food source, not only for the cod but also for Atlantic salmon. When the cod and salmon's food source is destroyed, they begin to eat their young. Atlantic salmon have evolved in such a way that they do not feed in rivers where their young are born, but it looks as if human overfishing is turning them into cannibals.

Loughlin says the cod and salmon stocks "will recover at the same rate as the restoration of the food chain." Dr. Wilfred Carter agrees. He points out that the overfishing of food stocks at sea has extended beyond the capelin fishery to sand eels and other tiny fish that support larger fish. "Hundreds of thousands of tonnes of those fish are taken every year, taken primarily to render them down into oil. Millions of small fish have been killed and converted into oil to heat electricity generating plants in Denmark. It's mind boggling that this kind of thing would happen, would be permitted to happen. There is still a pretty heavy exploitation of those fish food stocks, and I don't think that salmon are going to fully recover until there is a recovery in their food supply."

Peter Weber, a research associate with the Worldwatch Institute, an independent environmental research agency in Washington, D.C., writes in his 1994 assessment of conditions in the ocean that net fishing today destroys a lot more than the species of fish a boat intends to catch. "Coastal trawlers, which drag nets along the seabed, catch large quantities of unwanted species," he writes. Millions of tonnes of these unwanted fish are killed and discarded in the ocean each year.

The trawler damage extends to the ocean floor. "In the Dutch region of the North Sea, researchers have found that every square metre of the seabed is disturbed at least once a year by trawlers dragging nylon nets held open by beams and weighed down by chains. In the heavily trafficked areas, trawlers make an average of seven passes per year with these nets that weigh up to five tonnes each. The nets and chains dig into the seabed and kill sea urchins, starfish, worms, crustaceans, and shellfish, severely damaging the ecosystem."

Weber also makes note of the growing practice of "biomass fishing," where fishermen take all the fish they can with fine mesh trawling gear. "Much of this clean-sweep catch goes to meet the growing demand for feed for fish farming. In the most entropic possible disregard for the values of biological diversity, hundreds of interdependent species are ground up into a mash to feed a few monocultured fish. As fishers remove an ever greater proportion of the biomass from the marine environment, entire ecosystems begin to suffer."

Overfishing of the sand eel (a primary food source for Atlantic salmon) to produce fish meal and oil caused a collapse in the population of terns, puffins and other nesting birds in the Shetland Islands in the late 1980s. "The birds normally fed young sand eels to their chicks, but the fishes' population declined with the commercial catch, which peaked at 56,000 tonnes in 1982 and then slid to 4,800 tonnes in 1988," Weber writes.

"We can no longer afford to act as if the oceans are limitless or unalterable. The marine environment is integrally connected with the human environment and the biosphere, and there are few human industries or activities that do not ultimately affect the oceans in one way or another. The complex links between land and sea may make the task of protecting the oceans seem daunting, if not impossible. But it is precisely because of these links — because the oceans touch the lives of all of us — that we cannot ignore the health of the oceans if we are to protect our own place on the planet."

*

Along with the destruction of the ocean food chain, the temperature of the water in the North Atlantic is dropping, reducing the size of the feeding area for Atlantic salmon. Atlantic salmon prefer a sea temperature between 4°C and 8°C, and the area of ocean with water at that temperature is shrinking.

"My analogy is that, if you have a whole bunch of cows in a field, there's enough grass to start out, but if you start bringing the fences in, at some point you're not going to have enough grass and some of the cows are going to start to die," Anderson says. "And it appears that out there in the ocean, temperature actually may be a kind of fence."

Peterson and Friedland write that while there is a significant relationship between ocean temperature and salmon survival, it is not clear "whether this favourable or unfavourable habitat affects the salmon's food sources [or] its predation by other fish or seals, or if they just plain freeze to death."

"What can we do about it?" Anderson asks. "The answer is: not a damn thing. I mean, we're now talking about nature at work out there." But some scientists attribute this cooling of the ocean to global warming, which may be melting the polar ice cap too quickly, cooling the ocean's temperature and lowering the salinity of the water. A report from the United Nations Intergovernmental Panel on Climate Change, released in 1995, may back up this theory. The panel is now convinced that the release of greenhouse gases like carbon dioxide, caused by the burning of coal, oil and wood, is causing significant environmental changes.

Global warming may also be raising the temperature of water in our salmon rivers. Salmon live in fresh water at temperatures ranging from 0°C to 24°C; they spawn in temperatures ranging from 2°C to 8°C; and water between 27°C and 30°C can kill them. However, when water is 22°C, any additional stress can kill salmon. In July and August, 1995, water temperatures in some New Brunswick rivers ranged above 24°C. In extreme warm-water conditions, salmon seek out cold-water streams if they can and lie in those outflows until the main river cools down, but, during July, 1995, dead salmon were reported throughout the Restigouche watershed.

Carter points out that a program to reintroduce salmon into the Merrimack and Connecticut rivers, which lie at the southern edge of the Atlantic salmon's range, has been difficult. "They are not really making a lot of progress and I think their greatest problem is the temperature of the rivers," Carter says. "The rivers are simply too warm, particularly in the estuaries."

Over thousands of years the Atlantic salmon has gradually adapted to changes in water temperatures. In Ungava Bay, at the northern edge of the fish's range, salmon don't come back to their rivers until the middle of August because the mouths of the rivers haven't warmed up enough to allow them through alive. They have a two- or three-week window of opportunity to enter the rivers to spawn, but not enough time to get back out to sea. So they spawn and stay in their rivers until the following summer, when they leave during that three-week window. "They've adapted very beautifully to those harsh environmental conditions, but it's taken them eons to do that," Carter says. "We don't have those eons for salmon to adapt to warmer water. It's happening too quickly."

The summer of 1995 was the third-warmest summer in Canada this century, topped only by 1994 and 1989. Globally, the summer of 1995 was the warmest since records were first kept in 1866. "The spate of weather abnormalities during the past decade and a half has left experts speculating more and more if what we are seeing may be proof of climate change," said the Environment Canada report for 1995.

The day after Environment Canada released this report, the Toronto *Globe and Mail* editorialized about global warming, pointing out that there may be some positive results — shorter winters, a longer growing season and increased water from the melting of glaciers. "If global warming does become a challenge to human survival — and the case has yet to be proved — we will have to decide whether it is smarter in the end to fight or adapt."

At this point, Dr. Wilfred Carter would remind us of the miner's canary. In British coal mines, before there were sophisticated devices to measure methane gas, miners would take canaries underground in cages. When the canaries died, it was time to evacuate the people. "The salmon plays very much the same role in a way for our environment," he says. "If salmon for one reason or another leave a river that they have been returning to year after year for centuries, this is like the miner's canary, a signal that something has gone astray very badly and needs to be corrected." If Atlantic salmon are being driven from our rivers by global warming, these environmental changes may indeed be a challenge to human survival.

"In many salmon rivers, droughts and high summer temperatures are already a serious problem," Bruce Tufts, a fish physiologist in the biology department at Queen's University in Kingston, writes in the *Atlantic Salmon Journal*. "Even a fraction of the changes the climate models foresee could render useless all the restoration and enhancement efforts in the southern range of the species, and radically alter its distribution for generations to come. The more we learn about the biology of our planet, the more clear it becomes that the life of all the earth's species are interconnected in ways we never previously imagined. Atlantic salmon are a superb gauge of environmental health, but it's not just their fate that's at stake here."

*

The complex problems in the black box cannot be ignored, but at the same time we must nurture our troubled rivers and send as many young salmon to sea as possible.

Acidification of rivers because of acid rain kills salmon at the egg and fry stage, causing complete failure of spawning runs. Acid rain has destroyed salmon rivers all over the world, particularly in Nova Scotia, where more than a dozen rivers have completely lost their salmon runs. Norwegian rivers have been hit hard by acid rain. A group of acidified rivers in western Scotland is now fighting for survival.

In addition, agricultural pesticides poison rivers. "There's still a lot of those being used on New Brunswick agricultural land, and residues from those pesticides inevitably find their way into brooks and then into the rivers. Those are simply poison, not only to salmon themselves, particularly the small, juvenile salmon, but to salmon's food," says Wilfred Carter. Several years ago, Carter took a trip along the St. John River, looking for places to measure the impact of agricultural runoff. "We found there were little brooks running through the [potato] fields that the farmers actually used to wash their spraying equipment. They would bring the equipment down in the brook and hose it off, and all the residue would end up deposited in the brook. That was common practice."

Another threat to Atlantic salmon in rivers is disease and infection spread among wild stocks by fish farm escapees. And fish farming may present another, more subtle, threat to wild salmon stocks — the "genetic introgression" caused by he mating of wild salmon with aquaculture escapees. This is a major threat in Europe, where there are extensive fish-farming operations near most salmon rivers. In New Brunswick there are now tens of thousands of escaped farmed salmon in the Bay of Fundy, which is home to a smaller scale fish-farming industry. In 1995 on the Magaguadavic River, 89 per cent of the salmon run was made up of aquaculture escapees; of the 745 salmon captured in a trap on the river, only 79 were wild fish. That same year, the river received only 22 per cent of the eggs it needs to maintain its fish runs, and more than half of those eggs were from aquaculture salmon. Between 1992 and 1994, federal biologists took tissue samples from 150 fish and compared their DNA with samples from 75 salmon taken from the river in the 1970s, before the establishment of aquaculture in Passamaquoddy Bay. These biologists discovered "genetic compromise" in the Magaguadavic River salmon.

"The question is whether the interaction of escapees and wild fish leads to the progeny being less fit to survive in nature," says Anderson. "You can argue biologically that they might well be less fit, but there are geneticists who can argue quite forcefully that nature is quite capable of looking after herself. So the jury on that question is out. Suggesting that the aquaculture industry has had an effect is not an unreasonable working hypothesis. It's possible." One possible solution is for farms to raise sterile fish. But escaped sterile fish may become predators, eating smolt in the mouth of rivers. Another solution is for all fish farms to move to inshore facilities where the caged fish cannot escape.

In March, 1994, in a speech at the International Council for the Exploration of the Sea meeting in Copenhagen, Denmark, Dr. Kevin Friedland said:

> There is a wealth of poignant reminders of the importance of maintaining the genetic integrity of individual salmon stocks. The reminders are each unique river stock that has been lost in Europe and North America. What makes each stock so important are the local adaptations that are specific to that stock. These local adaptations are never considered until the resolve is developed to try and restore a stock that has been lost.
>
> There is a persistent concern over the effect that fish-farm escapees have on the genetics of stocks. Do these escapees interbreed, and what effect will such interbreeding have on stocks? The emerging answer is "yes" and the net effect is a reduction in genetic fitness in wild stocks. The future management of Atlantic salmon has been described as a conservation exercise as much as it is a fisheries management exercise.

It is clear that the emerging issues facing managers are new challenges to save stocks and habitat. The competitive use of salmon stocks and habitat will hopefully resolve itself in compromise because, as we have seen many times before, the process of recreating evolution (restoring a specific river stock of salmon) ranges from difficult to monumental.

The clearcutting debate

In 1979, in the watershed of New Brunswick's Nashwaak River, a group of scientists supervised the clearcutting of a huge tract of forest and then waited to see what would happen to the river. The experiment, called the Nashwaak Experimental Watershed Project, was carried out at a place called The Narrows, a well-known fishing area about 70 kilometres upstream from Fredericton. There, lumbermen clearcut 391 hectares of forest — an area four times the size of legal clearcuts in New Brunswick today. For the first one-and-a-half kilometres of the cut, they left no protective strip of trees along the river. For the rest of the cut, they left a 75-metre protective buffer of trees along the bank. The cutting was all done in the usual way forests were cut in the 1970s — with chainsaws and skidders. The logging slash was left on the ground and a control area of forest was left untouched by the scientists.

Dr. David Daugharty, a forest hydrologist and the assistant dean at the University of New Brunswick's Faculty of Forestry, was involved in the Nashwaak project from the beginning. He was part of a research team that included professors from the UNB forestry faculty and biology department, and scientists from the St. Anne-Nackawic forestry company, the New Brunswick government, the Canadian Forestry Service, the Canadian Wildlife Service and the Department of Fisheries and Oceans.

In the years after the cut, these scientists studied water levels in the stream, streamwater chemistry, and the soil, snow cover and vegetation of the former forest. The results? After every storm, some silt flowed into the river from the roads and stream crossings that were used in the clearcut. Water temperatures in the areas without the buffer strip were too high for salmon and trout. The flow of water in the river that ran through the cut area increased by 10 per cent.

Daugharty says the project's conclusions are straightforward. Poorly managed forestry operations harm rivers; well-managed cutting programs, clearcuts included, won't harm rivers. The Nashwaak project concluded that running machinery along the side of streams is harmful; that the less land converted into roads and landing areas the better; that properly constructed stream crossings are important; and that buffer strips along stream banks are essential. All of these issues can be addressed in a properly managed cutting operation. Most surprising is the study's conclusion that clearcutting does not cause low, warm water conditions in our rivers if a buffer strip of trees is left along the bank.

Poor road construction beside a river bank endangers salmon habitat, as seen here on the Dartmouth River, Gaspé Peninsula, Quebec.
ATLANTIC SALMON FEDERATION

Clearcuts are often blamed for causing water levels in our rivers to drop too quickly. Low water conditions in June and July are almost universally blamed on forestry operations. Daugharty says this is a myth. One of the reasons the Nashwaak project hasn't received a lot of attention is because it produced no dramatic negative results.

Jack Fenety says that in the 1950s and 1960s he was deeply concerned about extremely large clearcuts that came right down to riverbanks and endangered salmon habitats. He says we need to make sure cutting operations leave large buffers along our rivers. But in general he is less concerned than he used to be about what he sees happening in the woods today. "Clearcutting is not the great menace that most people would have us believe," he says.

Many conservationists have a different opinion of the effect of clearcutting on salmon. Environmental journalist Ted Williams writes in the *Atlantic Salmon Journal*: "With all the other ills besetting Atlantic salmon, forest removal has received little attention. No longer does one get the fine angling that came with slowly dropping water levels. The clearcutters have transformed entire watersheds from sponges that filter, cool and store water to dirt parking lots that pollute, heat and shunt it."

Dr. Wilfred Carter also expresses deep concerns about clearcutting:

> One of the worst things that we are doing and have been doing for some time in respect to salmon in particular is clearcutting. I think clearcutting is a curse, not only a blight on the landscape but a curse on the ability of rivers to reproduce stocks of fish. When I was a kid growing up in the Gaspé area, I can't ever remember years when, in the middle of June, you had a drought and some of the rivers were practically dry and the water temperatures were so high. And I think a lot of that problem is the direct result of clearcutting our forests. You lose the shade that the forests provide to the little feeder brooks and

> streams. You also lose the holding capacity of the root structure of the forest to prevent runoff. Once the snow melts, the rivers just flood right out and there isn't the massive root structure throughout the forest system that acts as sort of a sponge and holds the water back and releases it slowly through the whole season. That was nature's way of controlling the outflow of water and retaining water. Clearcutting also affects the production of tiny aquatic organisms that small fish eat. Changes of temperature in the water alter the composition of life in brooks and streams.

Carter's concerns should be taken seriously, especially his warnings about the effects of clearcutting on the natural balance of feeder brooks and streams. But his concerns about water levels may not be entirely justified. In the Nashwaak study, the overall flow of the stream after the clearcut *increased* by 10 per cent.

"Those trees use a lot of water," Daugharty says. "When you harvest, the regeneration that comes back doesn't use nearly as much. The thing we seem to forget is that when the rivers get low and warm, we haven't had rain for a month or so. When things get hot and dry it's because of some meteorological phenomenon, not because of some land-use phenomenon. Humans enjoy the hot, dry weather. They don't like the appearance of a clearcut. So when something has to be blamed, you pick the thing that you don't like, even if you're not right. With a clearcut you get more water, not less water, because those trees are after the same water we're after, and they get theirs first." Bad harvesting operations that allow skidders to scar the soil and create channels of water do drain land too quickly. A major network of roads with many drainage ditches can have the same effect, but Daugharty says this isn't a common problem today.

> The way a harvesting operation would be carried out today, there would be no increase in the efficiency of drainage in the basin. All you have to do is think a little bit about how that water makes its way down through the soil mantle, down the hill slope and eventually to the brook. That has not been substantially altered by the removal of the trees. That soil mantle is still of the same nature that it was previously. We go out and we casually observe these changes and we think we must have turned this into a very flashy basin with no opportunity for water storage, when, in actual fact, if you take measurements of the amount of soil moisture it's greater after a harvest than it was before.
>
> If there's any truth to this myth of a basin holding back water, that

> water is not going to percolate to the stream. That water is going to be used by vegetation long before it ever gets to a stream. If there's slower drainage, the trees are going to get it before the brook does, always.

Daugharty has extended his research into the effects of land use on river flows to a number of other important New Brunswick salmon rivers — the Little Southwest Miramichi, the Upsalquitch, the Restigouche and the Kedgwick. Data on some of these rivers go back to the early 1900s. "For all of those rivers we've gone back to the earliest period of record and tried to detect trends in stream flow and in snow melt runoff timing in the spring — trends that might be associated with change in land use over time — and we can't find them. They are not there. You can't find a land-use effect in the rivers. And yet folks of the Miramichi will say that they can tell the water goes up and down on their boots higher than it used to or faster than it used to. That's one of those things you can say, but I can't prove it with numbers and we'd sure like to know if that kind of thing can be found."

Fourteen years after the Nashwaak clearcut, water flows in the cut area and the control area were the same. Forests regenerate quickly, Daugharty says. The Nashwaak cut was not commercially replanted. But within a year, trees there were a metre high. After two years the trees were two metres high. When scientists returned to the site after three years, the trees were well above their heads.

*

Research programs like the Nashwaak watershed project are crucial sources of information to protect the future of our rivers. New Brunswick has a total of about 6.5 million hectares of land, half of which is Crown land managed by industrial forestry companies. These companies also control slightly less than 20 per cent of the rest of the province in freehold land. The remaining 30 per cent of New Brunswick is privately owned and includes a number of smaller, private woodlots.

Nine companies hold Crown timber licences in New Brunswick, and these companies pay the province a percentage of what they make from the wood they harvest off the land. All of the licence-holders are responsible for managing wildlife habitat, which includes salmon rivers. They must protect "riparian zones" along rivers, which essentially means no cutting within 30 metres of a watercourse. There are strict regulations about environmentally sound road construction and stream crossings. Deer wintering areas and moose calving grounds must be protected. Other environmentally sensitive areas such as those containing anything from an eagle's nest to rare plants must be protected.

The New Brunswick forest industry is big business. About 45 per cent of the province's exports are from the forests, and the industry provides 14,000 direct jobs and more than $400-million a year in salaries. The forest industry — and the clearcutting that goes along with it — is not going to go away.

John Gilbert, J.D. Irving Ltd.'s manager of fish and wildlife habitat, floats down the Downs Gulch stretch on the Restigouche River in northern New Brunswick. PHILIP LEE

The Irving group of companies is the biggest player in the New Brunswick forestry business. J.D. Irving Limited owns 950,000 hectares of freehold land and has management responsibilities on about one million acres of Crown land. John Gilbert says poorly built roads and stream crossings are a logging operation's greatest threat to salmon rivers. Most of the silt that fouls spawning grounds during a poorly managed cut is caused by runoff from roads. "If it is a good logging operation there should be zero silt," he says. Daugharty notes that road construction and watercourse crossings have improved dramatically in the past 15 years. The difference has been in the attitude of forest managers. "It's a matter of wanting to do it well," Daugharty says. "Operators have come a long way in the last 10 or 15 years in that regard." He says some companies today have specially trained crews that do nothing but watercourse crossings, and this has resulted in major improvements in our forests and streams.

J.D. Irving builds between 350 and 500 kilometres of new roads and installs between 1,200 and 1,500 culverts every year. All of these roads must be well planned, with proper runoff ditches, and well maintained, so water continues to flow into the ditches. Gilbert, who worked as a biologist for the province for 16 years before joining J.D. Irving, is responsible for making sure that the harvesting operations on these lands do not harm salmon waters and other wildlife habitat. K.C. Irving himself liked to go salmon fishing on those rare days when he wasn't working, and the sons and grandsons who now run his empire are salmon enthusiasts. "The owners like to fish and the owners live in New Brunswick, and as a consequence a lot of the things that we do in the woods reflect that," Gilbert says. "We are very conscious of doing a good job around the streams. We spend more money than anybody to make things right. We try to make sure that there's no effect from the forestry operations on aquatic habitat."

Forestry operations have changed since the 1960s and 1970s. The province limits clearcuts to no more than 100 hectares. The Irving operations take 60 hectares as their maximum, but their average cut size is more like 25 to 30 hectares, Gilbert says. These cuts are less than one-tenth the size of the cut in the Nashwaak project.

Trees used to be cut and skidded down a hill, creating furrows to carry water and silt down into rivers. Now tree harvesters with giant, soft tires run over a road of boughs and do not scar the soil. At stream crossings, road builders sow clover and hay to hold the soil in place. J.D. Irving Ltd. managers have occasionally told their stream-crossing staff to take photographs of culverts when they were in place. "You're not going to take a picture of something that looks crummy, because you have to take it back to the office and people are going to see it," Gilbert says. "You're making sure that it's the best-looking thing before you leave it." Gilbert says the law of the woods was laid down by J.K. Irving himself at a meeting of all woodlands superintendents. "He said anybody who's caught violating a law willfully because it's a benefit to the company is fired."

About 30 per cent of the company's harvesting now involves selective cutting. The rest is clearcutting. The company plants about 15 million trees a year on its freehold land, but some of the forests are left to regenerate naturally. Gilbert says the forests around Deersdale in the headwaters of the Miramichi have been worked hard by J.D. Irving, with a number of major cutting and replanting operations in the area. And he points out that the Miramichi's spawning tributaries there — Rocky Brook and Clearwater Brook — are still pristine.

❋

Forestry companies have been directly involved in managing salmon rivers. Fraser Inc. leases salmon waters from the province and operates fishing lodges for its guests. Avenor Inc. operates a camp at the mouth of Rocky Brook, an important spawning tributary on the Miramichi system, and owns a large section of riparian water on that river. "They have been great conservationists," says Fenety. "Rocky Brook is the greatest experimental laboratory for salmon in the world." An early run of salmon comes to Rocky Brook every year, filling the pools in May, but Avenor allows only limited fishing there. The Miramichi Salmon Association's satellite rearing program — where camp owners raise young salmon in streamside tanks — began at Avenor's camp in 1983.

Cattle grazing right down to the river bank cause erosion along the Nashwaak River, New Brunswick.
ATLANTIC SALMON FEDERATION

The Irving family owns or leases fishing rights on more salmon water than any other single group in the province. J.D. Irving Ltd. leases water from the province at Downs Gulch on the Restigouche River, owns fishing waters at Boston Brook on the Little Main Restigouche, and owns a large tract of land in the upper headwaters of the Miramichi River system.

In 1995, J.D. Irving Ltd. raised 41,000 salmon parr in two locations — at Boston Brook on the Little Main Restigouche and at the company's tree nursery at Juniper, in the Miramichi headwaters. In October, the company released 25,000 young salmon into Clearwater Brook with the help of students from Juniper. The company also has a deal with the high school in Juniper whereby the students will do the stocking at different points along the river. "It raises the whole awareness," Gilbert says. "The big thing is that when you've done something like stocking the stream, the chances are that you are going to respect it later on, and that you're not going to go out and net fish that you helped put in." In the spring of 1996, J.D. Irving Ltd. launched a major five-year research project with the Atlantic Salmon Federation to determine the maximum number of salmon that can be produced in Clearwater Brook and the Little Main Restigouche River. The scientific model developed on those rivers will be used to help develop salmon managment plans for all rivers.

*

The quality of forestry operations is by no means the only important factor in the protection of salmon habitat. Six years ago, a federal government study into habitat loss on Atlantic salmon rivers concluded that out of the approximately 18 per cent of Canada's salmon habitat that has been permanently lost in the last 120 years, 8 per cent has been lost to "chronic" land-use problems, primarily soil erosion on farms. A rolling, heavily grazed field that goes right down to the river bank is a beautiful sight but a menace to the river, causing erosion and warming of the water where there is no shade cover. (Poorly managed forestry operations would fit into this category, but weren't singled out in the study.) Dams and other obstructions — like the floodgates on New Brunswick's Petitcodiac River causeway that have destroyed that river's salmon runs, and the St. John River hydro dams (both specifically singled out in the study) — caused a habitat loss of 7 per cent. About 3 per cent of our salmon habitat has been lost to acid rain.

We need to fiercely protect and restore our salmon habitat. And we need to make sure that when we point fingers, they're pointed in the right directions.

Failing the test

On an October evening in 1992, I caught and released a 10-pound salmon at Secord Pool on the Kennebecasis River. Wet snowflakes were falling and dissolving on the surface of the water. Red and orange leaves were floating down through the pool. The river valley above the village of Norton was still and cold.

As soon as I released my fish back into the cold water and watched it swim back into the current with a hard thrust of its tail, Charlie Wanamaker, a Secord regular and obsessed salmon enthusiast who was fishing with my father and me that evening, hooked and released an even larger salmon. "It looks like we've got a run on here," he said. As dusk descended on the river, we left Wanamaker and went home happy.

The next year my father, Philip Lee Sr., and I went back to Secord Pool several times. We didn't even see a fish. The year after that the river was restricted to hook-and-release fishing only. The next year the river was closed. Our river hit rock-bottom fast and our little stretch of paradise, only a 45-minute drive from our offices in Saint John, was lost. Today, the future of the Kennebecasis River salmon runs is uncertain.

There are 550 Atlantic salmon rivers in eastern Canada. On these rivers, the federal biologists have set a target of 2.4 salmon eggs per square metre of spawning area as the minimum required to keep salmon runs healthy. They start with this number of eggs and then estimate the number of large and small salmon needed to produce them. These numbers of eggs and salmon should be seen as the bare minimum our rivers need, not as the long-term goal.

On rivers where there are no dams with fishways that allow fish to be accurately counted, biologists estimate numbers of spawning fish using a series of complicated mathematical models. On these rivers, the estimates often have a wide range — for example, a river might have 60 per cent of the eggs it needs, or 105 per cent, or somewhere in between. In 1994, federal government biologists conducted detailed assessments of 63 Atlantic salmon rivers. The results couldn't have been much worse. Forty-four out of 63 rivers failed the test. Only 19 rivers equalled or exceeded their minimum target number of eggs, and 24 rivers had less than 50 per cent of the eggs they needed.

Rivers that flow into the Bay of Fundy were found to have a serious shortage of eggs and returning spawners. The inner Bay of Fundy rivers (a series of small rivers east of the St. John) have been closed for all fishing since 1990. The returns on the Stewiacke River in Nova Scotia, which serves as an index for the inner Bay of Fundy rivers, have consistently been "less than conservation requirements." Salmon returns have fallen short by 80 per cent on the Stewiacke, and 72 per cent on New Brunswick's Big Salmon River. The outlook is poor. The number of salmon growing in the waters of the Stewiacke is at a historic low, and just 110 were counted in Big Salmon River in 1995. That exquisite river traditionally had runs of 5,000 salmon.

The outer Bay of Fundy rivers (west of the St. John River) aren't doing much better. In 1993, the St. John River had the lowest returns in 19 years — 4,369 grilse and 3,389 big salmon. Above the Mactaquac Dam, there were only about half the fish needed to seed the river. The next year was worse — 3,534 grilse and 2,375 big salmon — and 1995 saw a mere 7,249 fish returning to the base of the dam. The Nashwaak River, a tributary downriver from the Mactaquac Dam, received less than 40 per cent of the eggs it needed to seed the river in 1993 and just 30 per cent of the eggs it required in 1994.

The Restigouche River system needs 71.4 million eggs — about 12,200 spawning large salmon — each year. In 1993, it received between 28 and 52 per cent of the target number of eggs, and an estimated 3,200 to 6,000 large salmon returned. Returns were higher in 1994, when the river received between 56 and 101 per cent of its target number of eggs. But the number of large salmon is still about half the number needed to seed the river properly.

Biologists say that although the Nepisiguit River needs 9.6 million salmon eggs each year to maintain its fish runs, it received only 72 per cent of the target in 1993, and 59 per cent in 1994. (Some estimates of Nepisiguit returns are higher, but all are below the number required to keep the river healthy.)

The Miramichi River, the East Coast's brightest light, has met or exceeded its target of 132 million eggs in eight of the past 10 years. But there are still concerns about low numbers of large fish on this river system. And there are concerns about the health of some Miramichi tributaries, especially the famous Little Southwest Miramichi.

Smaller rivers are also in deep trouble. The Richibucto has been below its egg target since 1992, and the Buctouche River has been consistently below its target number of eggs. In the Petitcodiac River, salmon are on the verge of extinction.

*

Who is responsible for dealing with this grave situation? Canada's Constitution gives the federal government responsibility for "sea coast and inland fisheries." Meanwhile, the provinces have the power to govern the management, use and sale of water, land and minerals. Water use, the dumping of wastes, hydroelectric developments, roads, mining and forestry are all administered by the provinces and all have some effect on fisheries. Under common law, fishing rights in non-tidal waters can be private property and are administered by the province. Thus, a complex web of federal and provincial government agencies manage salmon. All Canadian provinces make agreements with Ottawa to help administer the federal government's fishery responsibilities. Whereas the federal government is responsible for Atlantic salmon conservation and management, the provincial government licenses fishermen and runs Crown Reserve salmon waters and fishing leases on our rivers.

Dr. John A. Ritter, the head of salmon conservation for the Department of Fisheries and Oceans in the Maritimes, says times are changing on our rivers: The federal government is seeking more help from river communities. "We still do have responsibility for the salmon resource. We will retain that. We are looking for people on the river to do more of the work on the resource, instead of leaving it to government, which has traditionally been the case. There are many benefits to that, benefits not only in terms of them actually doing the work, but also in terms of them obtaining a better understanding of the resource and how fragile it is and what the problems are."

Ritter says Ottawa is seeking more partnerships with local conservation groups and encouraging people on the rivers to raise money and take on projects themselves. "In some cases they'll have to recognize that we're also probably going to be like Big Brother in those activities, looking over their shoulder. But it goes beyond their just paying for certain things and actually doing them. It goes to actually accepting ownership of the resource and responsibility for making sure it's maintained."

At the same time, Ottawa is pulling out of salmon enhancement, most notably getting out of the fish-hatchery business. Ottawa plans to slash its annual Atlantic salmon river enhancement budget from $4.3-million a year to $1.7-million. In the future, federal biologists will be involved in river enhancement projects only as advisers to local conservation groups. Nine federal hatcheries in the Maritimes will be closed or privatized by March 1999 — four in New Brunswick, four in Nova Scotia and one in Prince Edward Island. In New Brunswick, Ottawa plans to stop running the Mactaquac Dam hatchery, which, like it or not, is the life-support system for the St. John River and which must continue operating until new solutions for that watershed are found. Ottawa will save $2-million a year when it abandons the hatcheries.

*

The man who conducted Canada's only comprehensive study of freshwater fisheries thinks this movement towards local management of natural resources is essential. And he sees local management groups having a lot more power than Ottawa has indicated it is willing to concede. In the vision of resource economist

Dr. Peter H. Pearse, Ottawa should set the guidelines for freshwater fisheries, and local management groups should take care of the resource. In the mid-1980s, Pearse was commissioned by the Canadian Wildlife Federation to conduct a thorough study of the country's freshwater fisheries. Pearse had the qualifications. He had conducted a major report on the British Columbia forest industry for the New Democratic Party government of Dave Barrett. And in 1980 he took on the challenge of studying the crisis in the wild salmon stocks on the Pacific coast, where a stock collapse had put thousands of people out of work and closed a dozen fish plants.

Journalist Geoff Meggs describes Pearse's strategy for the West Coast fishery as "revolutionary" (despite the fact that Meggs strongly disagrees with Pearse's conclusions). In *Salmon: The Decline of the British Columbia Fishery*, Meggs writes that Pearse wanted to transform the British Columbia salmon fishery from common property to private property. "The key to Pearse's proposals was a commitment to privatizing fishing rights," Meggs writes. Pearse said of the collapse of the B.C. salmon runs: "Stock depletion, poor economic performance and instability result from treating the resource as common property and are normal when resources are treated this way. It is the Tragedy of the Commons."

When Pearse turned his eye to Canada's freshwater fisheries eight years later, he produced a highly regarded but poorly circulated report entitled *Rising to the Challenge: A New Policy for Canada's Freshwater Fisheries*. His final report wasn't popular with some members of the Canadian Wildlife Federation because he suggested that users of our natural resources should pay for them, and he recommended that the management of fisheries be turned over to local non-government groups. He noted that this type of management could already be found on privately owned stretches of river in New Brunswick and Quebec.

> Although we need more management to control fishing and adapt supply to the real needs of users, we must be careful not to make this an excuse for heavy bureaucratic controls. Everything points to the desirability of managerial control by local user and community groups, a kind of local self-management, subject to ultimate governmental control and guidelines for the conservation of the fisheries resources.
>
> Local management is practised in Quebec and New Brunswick by private clubs and outfitters which normally regulate access to fishing waters in which they lease or own the fishing rights. Fishing rights and responsibilities could be assigned to community groups, local fishing clubs, Indian bands, and resort operators by means of licences that

> oblige them to manage the fishery according to a management plan annually revised and approved by the regulatory agency.
>
> Such agreements could relieve governments of the burden of direct administration and enforcement. They could bring non-government manpower and resources to the support of fisheries, encouraging participation, co-operation, and compliance with regulations. I am aware the very idea of user involvement in fisheries management can cause anxiety among some fishermen to whom the idea of private profit from public resources is anathema. It should not be embarked upon lightly. It is not a panacea for all fisheries, but rather a promising approach for dealing with fisheries under pressure.

Pearse also recommended that users help pay for managing the resource: "Licence fees are generally low, the right to fish for an entire season usually costing Canadians less than admission to one hockey game. In most jurisdictions the revenue from licence fees falls considerably short of the cost of managing the fisheries. I believe this is a major obstacle to improved fisheries management. Moreover, low-priced fishing encourages false perceptions of low-cost management and low-value resources.

"Valuable resources are at stake. Those who benefit from them are mainly fishermen. We who fish should be prepared to pay. Those who want to call the tune must pay the piper. This is not to say that society at large has no responsibility. Fisheries management goes beyond producing food and recreation; there are broader social interests in protecting fish and habitats, such as conserving heritage and environmental quality, and increasing scientific understanding. General taxpayers should contribute part of the cost of managing fisheries, as they already do. But fishermen will have to bear the cost of the largest part of the improved management we need today."

J.W. (Bud) Bird, New Brunswick's minister of natural resources from 1978 to 1982, a former Member of Parliament, and the owner of a salmon camp at Slate Island on the upper Miramichi, was elected president of the Miramichi Salmon Association in 1996. He says New Brunswick's unusual river management system, which mixes public water, private water, leased water and Crown Reserve stretches, "has served us pretty well in balancing out all of the interests, including conservation." But Bird thinks that a change in the way we manage our resources is coming. He agrees with Pearse that in the future local management groups may have to introduce some way of making people pay to fish on a river, and for this system to work, the money must go back into river management.

Bird is no stranger to salmon management reform. The Fredericton businessman was already a salmon conservationist when he came to the Natural Resources portfolio, and in four years he made a major contribution to conserving the resource. His most important reform was the controversial introduction of tags for all salmon. "It was, first of all, designed to eradicate the motivation for poaching by eliminating the black market. It had some tremendous extended values. All of a sudden we started to talk about salmon on a fish-by-fish basis instead of truckload by truckload."

Salmon conservationist Dr. Alex Bielak fishes Rocky Brook, a tributary of the Main Southwest Miramichi River.
ARLIE WYNN

Today, Bird thinks the development of local watershed management is essential to conserving our rivers. He sees local groups making decisions about guarding waters, setting season limits and enhancing river habitat. He says local groups could even run a river warden service. "Those could be paid for by the fishing public as government resources diminish. And that's going to pose a whole new regulatory challenge. If the Miramichi watershed is going to manage itself, where's it going to get its resources? Someone's got to pay to support the management, the conservation and the utilization of this resource."

Dr. Alex Bielak welcomes the idea of local conservation groups helping to manage our rivers. "We can't manage the whole continent. We can't say that what is good here is going to be good there. We need individual river management."

In New Brunswick there is already an example of the type of river management that Bird and Bielak advocate. Every 10 years the province leases stretches of our rivers to private individuals and clubs. They pay for the privilege of using the waters, and they must do the work to sustain the resource. As we struggle with the question of whether local management can work for natural resources, river leases are examples of private control of public resources that are already working well.

Tucker Cluett of Blue Hill, Maine, became the owner and leaseholder of a prime piece of Upsalquitch River real estate during a conversation with his friend Joe Sewall, a former state senator from Maine. The conversation went something like this: "All right, Tucker, I want to sell and I want to sell it to you. What do you think of that?" That was 15 years and several hundred thousand dollars ago. Cluett hasn't looked back since.

God never made a river finer than the Upsalquitch, an intimate, gin-clear, gravel-bottomed mountain stream that flows through hardwood hills and farm valleys into the Restigouche River. Cluett owns six acres of land along the river and leases the Grog Brook water from the Province of New Brunswick.

The 10-year Grog Brook lease cost Cluett $8,000 in 1989, with the payments rising by $400 a year. As part of the agreement he has to hire two wardens (one of them has to watch the water at night); he must create a 12-site campground on the Upsalquitch and hire two people to keep it clean; and he pays taxes on the leased water, which is assessed for about $120,000. Cluett is one of 19 parties with 10-year leases on salmon water in New Brunswick. By March 31, 2000, the province will have made more than $4.2-million from these stretches of river. And there are other financial benefits: the leaseholders hire a total of 89 resource workers every year who are not paid by the province.

Cluett's operation is not commercial. He runs the lodge for his friends and he fishes himself. The camp has three buildings: a comfortable lodge with a stone fireplace and two bedrooms, a kitchen and dining room with high ceilings and windows overlooking the river and a log cabin for himself. He employs three guides, three wardens, two riverbank and campsite cleanup people, a cook and a housekeeper. One of the guide's daughters serves afternoon tea. "I'm outnumbered," he says, noting that he often is the "chore boy" who goes to town to do the shopping.

Tucker Cluett's MacLennan Lodge is situated beside the clear waters of the Upsalquitch River. PHILIP LEE

MacLennan Lodge is open from the middle of June until the end of July, during the prime fishing weeks. All of this money and effort goes into six weeks of salmon fishing. "What I do is typical of the camps on this river system," he says. In some rural areas when the camps are closed, unemployment runs about 40 per cent; when the camps are open, the unemployment rate drops to 20 per cent.

Cluett is happy with his lease, but he notes that the campground and the two employees he hires for three months to keep it clean are a waste of time and money. After he won the lease, he constructed a campground with outhouses and picnic tables at a site beside the famous Crooked Rapids stretch of water. "We did that in 1990 and no one has ever used the campground. This is 1995. It was a make-work thing." Cluett abides by the terms of his lease, but his campground cleanup crew is clearing brush, painting and doing construction work on the camp.

The Grog Brook lease is prime salmon water that includes the wonderful Upsalquitch Home Pool. Peter Bodo, a senior writer with *Tennis Magazine*, a *New York Times* columnist and a regular guest of Cluett's, once wrote: "If the Upsalquitch were a golf course, Home would be the 18th hole at Augusta National. At the head of the 250-yard pool, the Upsalquitch rolls merrily over a prime wet-fly run and comes up against a steep bank and massive rock formation that forces the river to make a dogleg. The pool is very deep and slow at the turn, but then it flattens out, picks up some speed, and flows like subtly shifting glass over a series of ledges that extend out from the high bank for a good hundred yards. Typically, the fish lie in the boulder-strewn trough right at the abrupt edge of the ledges, in crystal-clear water between three and five feet deep."

❋

Fraser Inc., the timber giant, holds both the Lower and Upper Kedgwick leases and pays more per year for salmon river leases than any individual or company in the province. The Kedgwick is a fine salmon river that flows into the Restigouche. In the first year of its lease, Fraser paid $48,000 for the two stretches; the payment rose to $57,000 in 1995. In compliance with the terms of the lease, Fraser must construct and maintain a 12-site campground for canoeists and also hire seven river cleanup people and five wardens.

On the Restigouche, J.D. Irving Ltd. leases the beautiful Downs Gulch stretch of water, which runs from Upper Cross Point Pool down through a series of rapids and pools to the Tracy Pool at the mouth of Tracy Brook, where hundreds of salmon rest in the cold water. The classic Irving lodge, a former Ristigouche Salmon Club lodge, is located halfway down the stretch. (Ristigouche is an early archaic spelling of Restigouche.) Below the lodge is a straight, shallow stretch of river called the Trotting Grounds, named in the days when horses pulled barges upriver for sport fishermen. There are two constantly occupied wardens' cabins on the Downs Gulch stretch, one at each end of the leased water. Wayne Matthews, the manager of J.D. Irving's sporting camps, uses Downs Gulch as his headquarters in the summer. Matthews says this leased water is lightly fished and well guarded against poaching.

The J.D. Irving Ltd. camp on the Restigouche River's Downs Gulch stretch is one of the river's classic old fishing lodges. PHILIP LEE

Canada's most exclusive fishing club, the Ristigouche Salmon Club, based in Matapédia, Quebec, leases the Cross Point stretch of water on the Restigouche, which runs downstream from the famous Million Dollar Pool — a deep, swirling pool where the Patapédia River flows into the Restigouche. The Ristigouche Salmon Club has to build and maintain a 12-site campground and hire two cleanup people and four wardens.

Joseph Cullman III, the chairman emeritus of the billion-dollar conglomerate Philip Morris Inc., leases the renowned Two Brooks water on the Upsalquitch. In 1995, he paid $26,400 for the lease. Cullman, a longtime salmon conservation activist, operates a lodge at Two Brooks for his guests, and he, too, must build a 12-site campground and hire three cleanup people and four wardens. Cullman also operates Runnymede Lodge on the main Restigouche River. Edgar Cullman, his brother and partner at Runnymede, leases the water on the Restigouche in front of their lodge, the Rafting Ground Pool. The lease cost $7,200 in 1995 and a warden must be hired to watch the pool. "We handle the waters we lease as if we owned them," Joseph Cullman says.

Goelet Estate Company, a company that owns a large tract of land in midtown Manhattan and is itself owned by Francis Goelet, the chairman of the Goelet Corporation, holds the Toad Brook lease on the Restigouche River. Goelet is a salmon conservationist who has given hundreds of thousands, if not millions, of dollars to the cause. His family fishes the Restigouche, while he spends most of his time at his lodge, called Tracadie, on the Grande Cascapédia River in Quebec.

Four New Brunswick fishing clubs also hold leases: Tabusintac Club Ltd. (Tabusintac River), Square Forks Fishing Club (Big Sevogle River), Thistle Fishing Club (Little Southwest Miramichi River) and Aesculapius Fishing Association (Lower North Branch of the Little Southwest Miramichi). The historic Miramichi Fish and Game Club, which has six American and four Canadian members (including Jack Fenety) leases water on the Northwest Miramichi River; the club paid $12,600 for its lease in 1995.

*

Almost all of these leased waters are lightly fished. Cluett shares his waters with a downstream camp called Millbrook Farm, operated by his friends Sewall and newspaper publisher Rick Warren, fellow residents of Maine. Cluett first came to the Upsalquitch when Sewall was running Millbrook Farm commercially and had purchased MacLennan Lodge in order to have more water for his guests to fish. Then, because he wanted to get out of the outfitting business, Sewall decided to sell MacLennan Lodge to Cluett so they could share the water.

Cluett fishes three boats and has four guests at his lodge at a time. The fishing over the past 15 years has generally become a little bit better, he says, although he can't account for some of the wild variations — 1993 was one of his worst years; 1994 was his best year ever; 1995 was terrible and he closed the lodge early. His fishermen landed 232 fish in 1994; the best year before that was 180 fish. "You think it would get steadily better and better as the [ocean] netting decreases, but there are other factors," he says.

Cluett, like the other leaseholders, monitors his own pools. In 15 years no government official has ever asked one of his guests for a licence or come in to check to see if the fish in the ice house are tagged. Once every two weeks a federal warden drops by. When you fish one of Cluett's pools, you don't see anybody else; you've got the whole pool to yourself. "The [number of] fish we catch compared to the fish we see is minuscule," he says.

He has first-rate guides, most notably brothers Shane and David Mann. Cluett depends on them to find fish for his guests. He loves it when his guests catch fish and gets depressed when the fishing is poor. When Cluett closes his lodge at the end of July, the Mann brothers look after his waters and Cluett permits some local people to fish. People who fish the water in August are asked to report all the fish they take because Cluett carefully accounts for every fish hooked in his stretch of river.

These records are valuable for scientific reasons, and they are also useful in his friendly but fierce rivalry with Sewall and Warren. One afternoon in July, 1995,

J.D. Irving Ltd. guide Emerson Main motors up toward Upper Cross Point Pool on the Restigouche River's Downs Gulch stretch. PHILIP LEE

John Albright, the director of the Atlantic Salmon Federation's operations in the United States, was sitting in Cluett's screen porch at MacLennan Lodge chatting about the river. Albright was a guest at Sewall's Millbrook Farm, and he casually commented that MacLennan Lodge was running ahead of Millbrook Farm in the fish count this year. "We've been running ahead of Millbrook Farm for 15 years," Cluett replied with a smile.

CHAPTER FOUR

Dam It All

Dr. George Frederick Clarke

No salmon leaped to fall back and send up a shower of diamond spray. Desolation broods over the once-lovely scene. Only the encircling hills are the same — silent as of old.

— Dr. George Frederick Clarke

History is long and our memories are short. What do we remember about the glorious salmon runs on our mighty St. John River, the longest river between the St. Lawrence and the Mississippi? A dinner hosted by the New Brunswick Salmon Council at the end of the 1995 fishing season opened with an announcement of "good news" for the St. John River — 7,249 salmon and grilse had returned to the base of the Mactaquac Dam. Compared to the 5,519 fish that returned in 1994, it was indeed good news. But in another context, this well-intentioned announcement could have just as easily been met with outrage and scorn.

Michael Brislain is the author of *Bugging the Atlantic Salmon*, a study of fly fishing on the St. John River. He points out that the river once had hundreds of thousands of fish running upstream to spawn before it was choked with pollution and dammed up. Brislain writes: "At the turn of the century, 100,000 salmon were commonly landed as part of the commercial catch in the harbour of Saint John. This huge catch, combined with catches from other Bay of Fundy rivers, made up the millions of pounds of salmon destined for export to the markets of New York and Boston. Every freehold adjacent to the river strung gill-nets and set nets to supplement the farm larder. Salmon were so common that the great fish was used for everything from an export commodity to fertilizer. Today, a total annual run of 10,000 fish, of which 4,000 are large salmon, is considered good." The run of 1980 was considered to be "the great run" — 10,000 large salmon and 17,000 grilse returned to the Mactaquac Dam.

To understand the significance of what has happened to the St. John, we must understand its history. No one understood the history of the St. John better than

Dr. George Frederick Clarke was one of New Brunswick's great conservationists.
TELEGRAPH JOURNAL

conservationist and author Dr. George Frederick Clarke, who was born beside the river in 1883. He fought with word and pen to save it from the three hydroelectric dams that would destroy it. On a cold winter night in the early 1960s, New Brunswick poet Alden Nowlan, who at that time was a reporter and editor for the *Telegraph Journal*, covered a public meeting called to protest the Mactaquac dam project. Years later, in an essay about his time at the newspaper, Nowlan recalled that night, how he drove on icy gravel roads and called in his story to the newspaper from an isolated telephone booth beside a gas station that had closed for the night. And he recalled how one man in particular spoke passionately against the dam and the "flooding of farmlands under cultivation, sometimes by the same family, since the arrival of the Loyalists."

Nowlan wrote: "The speakers included Dr. George Frederick Clarke, then in his eighties, a Canadian man of letters in the old tradition: he had contributed short stories to *Blackwood's Magazine* and could build a birchbark canoe. That night he called on the farmers to meet the representatives of the New Brunswick Electric Power Commission in the manner of their forefathers, the King's American Dragoons, once the finest cavalry regiment in the world. 'Answer them with your rifles,' the old man said. The farmers and their wives grinned and avoided one another's eyes."

Despite this call to arms, Clarke was a peaceful man, who understood the importance of wild rivers and was searching for a better way to live. "Every man, whatsoever his circumstances, is equally poor who fails to realize not only the beauty, the goodness, the kindness, and the charity that surrounds us, but also that he may be the master of his inner self. Although I cannot always retain that tranquility of spirit which is the ideal of Nature, I can at least think of it and try to attain it."

*

The St. John is a giant watershed. The river and its tributaries, rising from headwaters in Maine, drain more than 67,000 square kilometres of land, an area larger than Nova Scotia. Every county in New Brunswick, with the exception of Gloucester, has waters that flow into the St. John system.

However, the St. John has been badly damaged by pollution. Lumber mills, tanneries and starch plants have dumped poisons into the water. Farms and agricultural processing plants have filled the river with organic wastes, insecticides, herbicides and fertilizers. The forest industry and pulp and paper mills have deposited organic wastes and chemicals into the water. In recent years the river has been cleaned up, although it still has a way to go, especially in its lower reaches. Saint John Harbour is highly polluted, and the pollution runs upstream daily with the Bay of Fundy's powerful tides.

Dr. George Frederick Clarke with two days' catch at Gulquac Lodge, Tobique River, 1927. PANB

And then there are the dams. The construction of the Tinker Dam on the Aroostook River in 1922 eliminated the Atlantic salmon from that productive tributary. In the 1950s, the dam builders went to work on the Tobique and St. John at Tobique Narrows and Beechwood. The dams blocked the natural fish passage upstream, and the headponds destroyed large spawning areas and created collection ponds for pollution. Smolt travelling downstream through the dam turbines died in large numbers when they were battered by the turbines.

Clarke writes that the St. John was the most beautiful river in North America "before the dams at Beechwood and Mactaquac destroyed it for all time. The old river, which de la Galissonières and de Vaudreuil declared to be one of the most important gateways to New France, and tried so hard to hold, no longer tumbles with a song on its lips over the bars and around the heads of its myriad islands. From Mactaquac to within four miles of Hartland, or a distance of 56 miles, it is now a lifeless pond, and no more do salmon and grilse speed up it in the thousands to deposit their golden eggs in the far waters of the Tobique; and what once were gravelly bars in the St. John itself are now covered with noxious silt and pollution. Islands and intervales are now far beneath the headpond."

The pool at Hartland was once one of the finest salmon pools in the world. At the upper end of the town, the St. John is joined by Becaguimec Stream, a cold-water source that lowers the temperature of the main river. A cold spring also bubbles up in the middle of the pool. Clarke said: "Although the fish were taken here soon after the spring freshet subsided, the best angling was from the first of July to the last of August, when the fish found this cooler water in astonishing numbers, and so close in did they lie they could be successfully cast over from the beach a few yards out."

He describes the scene one July when there was an unusually heavy run of salmon in the St. John, and hundreds of salmon sought refuge in the cool waters of the Hartland Pool. "They began taking shortly after 6 p.m. Saturday evening. In a short time the shore was lined with anglers. By Sunday night, or in about 15 hours of angling, 150 salmon and grilse had been hooked, played and landed. I venture to say no pool on the Restigouche, nor any other in the world, has ever produced so many fish in the same length of time."

Clarke noted that these Sunday anglers were the targets of criticism from the local clergy and travelling evangelists. "It remained for the New Brunswick Electric Power Commission and the dam at Mactaquac, with its 60-mile headpond, to accomplish what prayers and anathemas had failed to do. The door was shut. Howls of indignation! So, in September, 1969, the Fisheries Department dumped a few salmon of grilse-age into the pool. In diminished numbers, fishing was resumed. A few were caught. But, as someone put it, it was like hooking pet goldfish."

Before the construction of the Mactaquac Dam, salmon used to return to the base of the Beechwood Dam, which blocked the river upstream from the Hartland Pool. Clarke said that one year 4,000 fish were tagged and lifted up over the dam in huge nets. "But seemingly they didn't like the water of the strange headpond or were bewildered at not finding running water, for they sought the outflow. The turbines must have caused a mortality of 30 per cent. But others got through intact, and many were caught, metal tags and all, by the fishermen. Beechwood had closed the door to the Tobique waters."

*

Clarke understood the St. John River through its people. He writes of a friend of the river named Russell Boyer, a native of the river town of Victoria Corner, who spent his life working as a guide and warden. After Boyer's wife died, he spent the last five summers of his life living on a simple campsite on an island in the river near his home town. Boyer loved to receive visitors on his island, and Clarke would break away from his office, travel to Victoria Corner and shout across the river to his friend; Boyer would appear in his canoe to ferry Clarke across to the campsite. The two men would sit and drink and talk for hours, telling river stories and reciting poetry. At night, Boyer often ran the river in his canoe, watching for salmon poachers. They would never hear him coming, only his voice telling them: "You better come to shore, boys." Boyer died before the dams came to his river.

Clarke writes: "In 1950, Russell was laid to rest in the little cemetery that overlooks the St. John and the island that he loved so well."

*

Clarke believed in rivers and their people, and he always turned to rivers and the wilderness to restore his soul. "A day off is a day gained," he wrote. "The pity is that we do not snatch enough of them from the years. It matters not if our day off be beside a wilderness lake or a little river, or the seashore, or wherever fancy takes us, so long as it renews and satisfies our spiritual hunger and helps us the better to face life with its numerous complexities, its joys and sorrows."

Fishermen gill-net salmon in Saint John Harbour. This fishery is now closed.
ATLANTIC SALMON FEDERATION

His fine writing and conservation ethic, in the end, were part of his search to understand the nature of human virtue. Clarke's "other river," in *Six Salmon Rivers and Another*, is the river of this search.

The Tobique Salmon Club

Take a drive through the village of Matapédia, Quebec, and follow the gravel road that snakes along the river. At the end of that road, you'll find a piece of New Brunswick history — the Tobique Salmon Club on the Matapédia River. Most of the people associated with the club today don't know why this prominent Quebec salmon club bears a New Brunswick river's name, a river named after Tobit, a Maliseet Indian who once lived near where the Tobique flows into the St. John. They don't know that the Tobique Salmon Club was driven from a New Brunswick salmon river by the construction of hydroelectric dams. They don't know that the Tobique River was once part of one of the world's greatest salmon river systems.

The Tobique Salmon Club was formed on the St. John River tributary in 1889. The club hosted 12 guests at a time and owned and leased pools along 80 kilometres of the river. It had large camps at the Tobique Forks and Riley Brook and smaller camps at Blue Mountain Pools and Two Brooks Pools. The salmon fishing was wonderful. Today, the remains of the huts that housed river guardians can still be seen along the river at three-mile intervals, and a painting of the old Tobique River camps, with guides poling canoes along the river, hangs over the stone fireplace in the lodge on the Matapédia.

It all started to fall apart in 1953, when the New Brunswick Electric Power Commission built a dam at Tobique Narrows. In 1957, the commission built another dam, this time on the St. John River at Beechwood. By then, a determined man named Harold D. Hodgkinson had seen enough. Hodgkinson, a department store owner from Boston, was president of the Tobique Salmon Club. He believed the club's riparian rights on the river had been violated when the wild fish runs were blocked by the dams. The power commission had offered nothing in compensation, so in January, 1962, Hodgkinson turned to the courts. He hired lawyer George Baird, of Perth, who filed a claim with the New Brunswick Court of Queen's bench demanding $421,000 to compensate for the destruction of the club's salmon fishery.

The club brought its catch records to the court, and these carefully maintained log books told the story of the demise of the Tobique salmon runs. Between 1922 and 1931, club members caught 3,915 salmon. Between 1932 and 1941, the Depression years, when fewer club members travelled to New Brunswick, the camps recorded 2,418 fish killed. Between 1942 and 1951, club members caught 4,394 fish, recording a high of 626 in 1951. Between 1952 and 1961, the club caught just 2,980 salmon, despite a record 783 fish in 1953. The fishery bottomed out in 1957 when just four fish were caught. The records show 57 salmon caught in 1961. The federal government's fish counts were also entered as evidence in the case and they showed a steady decline in the number of salmon passing through the fishway at the Tobique Dam — the fish run at the dam declined from 4,656 in 1953 to 747 in 1961.

In response to Baird's claim, the power commission denied all responsibility and insisted that the dams didn't block the passage of spawning salmon migrating to the Tobique spawning grounds. Nor did the dams obstruct or impede "the progress of young fish and smolt downstream to salt water."

In June, 1964, Mr. Justice C.J. Jones ruled in favour of the fishing club. He set the tone for his decision when he declared that the Tobique River "is or formerly was one of the finest salmon fishing rivers on the continent." The judge said to establish "injurious affection" the club simply had to show that the dams had damaged its property "irrespective of negligence." He wrote: "A great deal of evidence was adduced to show that since the dams were built the fishing on the Tobique had deteriorated and was almost a thing of the past. These witnesses were mostly not disinterested but of a very impressive type and I have no doubt that their testimony was true. Does this condition exist as a result of the construction of the dams? Frankly, it would be straining the bounds of credulity too far to believe otherwise." The judge noted that a new pool had been created below the Beechwood Dam that had proved to be a fish-killing "bonanza" for anglers. Those fish, he added, should have been migrating up into the Tobique River system. "I am satisfied that it was an injurious affection to the property."

The judge also noted that between 1953 and 1958, the forest in the Tobique River valley was sprayed with DDT to fight the spruce budworm, and that this spray program was killing salmon. In an ironic moment during the trial, the power commission had argued that the spray program was the cause of the loss of the salmon fishery — this was a case of one arm of government blaming another for an environmental disaster. "There is no doubt in my mind that it [the spray program] had something to do with it but to what extent it is impossible on the evidence for me to say," the judge concluded.

The power commission appealed Jones's decision and lost, and in April, 1967, the courts asked that the amount of damages be settled by arbitration. By that time the power commission was about to deal the Tobique River a final crushing blow — a giant $110-million dam on the St. John at Mactaquac. The power commission and the Tobique club settled their case for $400,000. The club closed down and the riparian water, for all it was worth then, went back into the province's control.

Henry Streeter, the current president of the Tobique Salmon Club, says the late Hodgkinson was "a wonderful man, but if he got riled, he was a tough fighter." The club members got into a fierce debate over what to do with the $400,000. "There was a feeling that [since] the membership was getting along in age, why not split the pot and go home? But Mr. Hodgkinson said, 'No, we're trustees for the next generation of people coming along. We are going to use this money for fishing.'" In the late 1960s, Hodgkinson bought a camp on the Matapédia River that included an extensive tract of riparian water. Hodgkinson called it the Tobique Salmon Club and recruited younger members, including Streeter, who joined in 1971. Streeter, a 75-year-old native of Boston, is now recruiting younger members himself.

*

Ned McCaskill, a former Tobique Salmon Club guide, says there are few fish in his river now. When the Tobique Dam and fishway were put in, the salmon were able to pass over the dam and they had great fishing that first year. But the power commission didn't put in fishways at Beechwood or Mactaquac. The Beechwood Dam's fish lift "hardly worked at all" in its early days, McCaskill says. "Everybody used to kick about it but it didn't do any good. I guess they were just experimenting. They wouldn't listen to anybody."

McCaskill guided for years on the Serpentine River, a tributary of the Tobique that flows out of the Serpentine Lakes. "We had wonderful fishing on that river," he says. Everybody at Riley Brook was either a warden or a guide, or worked in some way for the sporting camps. "We had never seen hard times in the stock market crash in the thirties. We always had sports, and the rich people would come for the fishing. Now there's none of that, it's all changed. It kind of ruined our living for a while. It would almost make you cry to see the river there with no fish. I don't ever expect them to come back."

*

The Tobique has some of the finest salmon spawning grounds anywhere. The trouble now is that many of the fish that are spawned in the Tobique system die in dam turbines when they try to swim to sea. And the fish born in the giant Mactaquac hatchery want to return to hatchery waters; when they are dumped upstream in the Tobique system, they become confused.

"There's a lot of people who have reservations about whether the hatchery actually contributes up there," says Danny Bird, a former Tobique River guide who is now a regional co-ordinator with the Atlantic Salmon Federation. "For years, there were a lot of fish that hung around the Tobique Dam and they went back downriver from the dump site, probably trying to get over the dam and back down to Mactaquac if they could. There was one instance when they dumped a load of hatchery grilse into the Tobique and the locals actually caught them going back downriver."

Bird, a former president of the Tobique Salmon Protective Association, has worked for years to try to preserve what is left of the salmon runs on that river. "I realized that there was a possibility that we might be able to rebuild the resource. It was world class at one time and I believe that there's a way that that can be realized again."

Bird says conservation groups, the provincial and federal governments and N.B. Power have worked hard over the years to improve the salmon lift at Beechwood and the fish pass at the Tobique Dam. They now know they can get fish upstream around the dams. "We have to address the downstream problem. We're working on it every day. This river was once the Rolls Royce of salmon rivers."

The Serpentine Run

There is a good chance [the Mactaquac Dam]
may even increase the [salmon] runs.
— H.J. Robichaud

The St. John River's most valuable and rare genetic strain of Atlantic salmon — the Serpentine Run — is now measured in mere dozens. The story of the Serpentine Run reveals what has happened to all of the St. John's salmon runs. And if the Serpentine Run can be restored, so may all of the wild salmon runs on this great river system.

The Serpentine Run salmon enter the St. John River late in the fall, shortly before the river freezes, after spending about a year and a half at sea. Most salmon swim into rivers in the spring, summer and early fall and spawn right away. The Serpentine fish, on the other hand, don't spawn during their first fall in fresh water. They spend the winter under the ice in the lower reaches of the river, resting where the flows are gentle, probably travelling no further upstream than the wide, deep waters of Long Reach, just north of Saint John. Before the construction of the hydroelectric dams at Mactaquac, Beechwood and Tobique Narrows, thousands of these salmon would swim upriver in the spring and arrive at the Tobique River and its most important spawning tributary, the Serpentine, by May. This unique migration pattern may have developed because of the size and length of the St. John River and the distance these salmon needed to travel to reach their spawning grounds. They are larger than a grilse but smaller than most other salmon, "about eight pounds, short, just like bullets. This could be one of the purest genetic strains of early-run fish in the world," Danny Bird says. "It would be a shame to see them go extinct," agrees Jim McCaskill, the manager of the Mactaquac Hatchery, who has been carefully trying to preserve the few remaining Serpentine fish. Biologists at the hatchery managed to collect 20 of them in 1995 to use for brood stock.

The St. John River was once one of the greatest Atlantic salmon rivers on the continent.
ATLANTIC SALMON FEDERATION

In the spring of 1996, 25 Serpentine fish showed up at the dam; 22 were kept for brood stock and three were released upstream to spawn.

In the early 1970s, after the construction

Anglers once lined the banks of the St. John River before the three hydroelectric dams destroyed the river's legendary salmon runs.
ATLANTIC SALMON FEDERATION

of the dams, the Department of Fisheries and Oceans operated a trap at Westfield, just below Long Reach on the St. John. There, Serpentine fish were caught, tagged and released when they entered the river in October and November. The tagged Serpentines turned up at the base of the Mactaquac Dam in May.

N.B. Power built an $18.6-million hatchery at the base of the dam to compensate for the heavy smolt losses caused by the turbines at the dams. The federal government has been operating that hatchery ever since, at a cost of about $350,000 a year. Some argue that the Mactaquac hatchery project was indeed successful for a number of years; they blame the recent decline in numbers on the drastic reduction in salmon survival at sea. They point out that until recently, the river and hatchery often produced its target of 10,000 to 20,000 salmon a year. But how and why did they arrive at such a low target number for salmon in this huge watershed?

There's no question that the St. John runs were beaten down by agricultural and industrial pollution and overfishing that had nothing to do with the dams. But the dams have played a central role in the decline, despite the fact that the Mactaquac Hatchery has put more fish into the river. The federal government's own surveys on the river are cause for alarm. Since 1985, the spawning grounds above Mactaquac have failed to receive the minimum number of salmon eggs. In 1993, the numbers of eggs on the spawning grounds were at the lowest level in 20 years. In 1994, the upper reaches of the river received just 39 per cent of the egg target set by federal scientists.

*

Governments claimed that the hatchery could replace nature's intricate genetic work. The near-disappearance of the Serpentine Run shows that they were wrong. "The main contributor to the decline of the salmon run on the St. John River would be the near-disappearance of the Serpentine Run of fish," says Danny Bird.

In order to restore the Serpentine Run on the St. John, young fish, spawned in the upper reaches of the system, must get to the sea alive. Upstream passage for salmon used to be a problem, but this has been largely solved by a trucking system that moves fish above Mactaquac, an improved trap-and-lift at Beechwood and a functioning fishway at Tobique. Certainly, like many other rivers, the St. John is suffering because large numbers of its young fish are dying at sea, and until that turns around, returns to the river will continue to be low. However, problems at sea are no excuse for inaction on the river. Safe smolt passage downstream is essential to provide some hope for the future of the St. John's wild salmon runs.

In October, 1965, then federal minister of fisheries H.J. Robichaud made a detailed statement about the Mactaquac Dam and the future of St. John River salmon that was published in the *Atlantic Advocate*. Defending himself against charges that he had been delinquent for not opposing the dam, which conservationists had predicted would destroy the salmon, the minister said that, while it was the province's decision whether or not to build the dam, it was clearly his own responsibility to ensure that "populations of valuable fish are not unduly interfered with by any structure." Robichaud said his main concern about the project was how it would affect the Atlantic salmon: "Most St. John River salmon spawn above the Mactaquac site and the denial of these upriver spawning beds and rearing areas would be a catastrophe."

Before Mactaquac, many of the smolt spawned above the dams at Beechwood and Tobique Narrows came over the dams in the "spill" of water caused by heavy flows in the spring. Smolt going over the dams in the spill survived much better than they did when they ran the turbines. However, at Mactaquac there would be little or no spill. Robichaud said his scientists "came to the conclusion that the total turbine mortality from the three dams would be about 50 per cent of the smolt runs."

The Mactaquac Dam and its headpond have contributed to the rapid decline of the St. John River salmon.
ATLANTIC SALMON FEDERATION

The scientists also predicted that if headwater storage in the future reduced or eliminated spill at Beechwood Dam and forced all smolt through its turbines, smolt mortality would be even worse. "Nature has a wonderful way of compensating for interference with the perpetuation of a species, but not at the level of 50 or 60 per cent reduction," Robichaud wrote. "It was concluded, therefore, that the eventual result of turbine mortality of this size would be virtual elimination of the runs of salmon."

As a solution, Robichaud said he had decided to approve "somewhat unusual fish facilities which have been effectively introduced in Sweden." The facilities may have been effectively introduced in Sweden, but never on a river system the size of the St. John, and never on a river where fish managers were trying to protect wild fish. In Sweden, the technique of building a hatchery at the base of a dam was used on rivers where wild salmon runs had been completely destroyed. There, the hatchery was used at the base of the dam to maintain a commercial fishery. This method of maintaining salmon runs above a dam, in fact above three dams, was entirely experimental and has proven to be a failure.

How could the federal government justify using such an experimental program on the St. John when the federal minister of fisheries publicly admitted that the failure of the program would mean "the virtual elimination" of the runs of salmon? "I believe it is a good solution with a good chance of complete success in maintaining the St. John River runs of salmon undiminished," Robichaud said. "There is a good chance it may even increase the runs." The hatchery planned to release 500,000 smolt a year. On the upper reaches of the river, the experiment has largely been a failure. It's one thing to get 10,000 salmon to return to a hatchery. It's quite another to perpetuate the salmon's complex life cycle "undiminished."

The plight of the Serpentine Run underscores the problems of the Mactaquac hatchery system as a whole. Each St. John tributary, from the Tobique and Serpentine to every tiny spawning stream, used to have its own genetic strain of salmon that returned to the very area of their birth to spawn. Over thousands of years these salmon substocks had developed numerous adaptations to the conditions within their tributaries. For example, the shapes of their fins may have developed subtle differences to help them move through their home waters. When these fish return to the Mactaquac Hatchery, they interbreed. Dr. Larry Marshall, a senior Department of Fisheries and Oceans biologist who monitors the St. John River, says, "I do have concerns about what a hatchery can do towards homogenizing a number of distinct stocks. The one stock that has been most visible and most easily identified is the Serpentine Run. But substocks are not always as physically distinct as that Serpentine Run is. I think it would be a fair assumption that there are probably other substocks that we can't identify with the naked eye."

Another problem is that hatchery-born fish go to sea and return to their home waters to spawn. At the dam, biologists can tell the difference between hatchery fish and wild fish: hatchery fish usually have dorsal fin deformities that develop when they are young fish in tanks. Biologists make sure that most of the wild fish are released above the dam, if possible. When hatchery fish are trucked around the dam, they find themselves in foreign water. They don't know where they are going because their homing instinct is to the hatchery water. "How does that hatchery fish know not to spawn on the main stem of the St. John?" Marshall asks. Eggs deposited in the fluctuating waters of the main stem of the river have less chance of surviving than those buried in nests in the safe waters of the upper Tobique.

Dr. Wilfred Carter says the Mactaquac Hatchery has contributed a lot of additional salmon to the river, but such systems are really only useful as interim measures. "I don't think it's desirable to try and maintain a natural population of salmon by totally artificial means, by hatcheries alone," he says. "So a hatchery is a crutch that you need to use temporarily while you have a problem. But I don't think you could ever hope to use a hatchery to substitute for natural production."

The staff at the hatchery deserve credit for saving the Serpentine fish. Jim McCaskill, the manager, took steps to save this stock from extinction long before the federal government was interested in the Serpentine Run. For years, the Serpentine brood stock have been separated from other strains and bred only among themselves. Some young Serpentine fish are bred and transported upstream; others are released from the hatchery as smolt. The hatchery breeds about 30,000 young Serpentine fish every year and puts them in the Serpentine River to grow and eventually run to sea as smolt: Nevertheless they still suffer heavy losses as they swim through the turbines, and only a dozen or so make it back to the dam every year.

*

There are no easy solutions to the problem of getting smolt past the turbines alive. C. Paul Ruggles, a retired Department of Fisheries and Oceans biologist who is now a consultant specializing in fish and hydroelectric projects, was involved in the negotiations with the power commission that resulted in the construction of the Mactaquac Hatchery. He knows the St. John River system well, and he also knows about efforts to save salmon runs elsewhere.

In the past 40 years, $2-billion has been spent on the Columbia River to try to find a way to keep salmon smolt alive on the way downstream. "This is like going to the moon," Ruggles says. The Columbia, which flows through Oregon, Washington and Idaho — and eight massive hydroelectric dams — is still losing its salmon runs. The most recent structure that was built on the Columbia to try to keep the smolt away from the turbines is killing more fish than the turbines did — "killing them with kindness," Ruggles says.

On smaller rivers, scientists have tried using lights, sounds and electrical impulses to try to draw fish away from the turbines. None of these methods works. Screening devices can be used to divert smolt from turbines, but these require a narrow water passage leading to the dam, which is not present at Mactaquac. Ruggles says there may be no structure that can be built on the St. John that will effectively divert smolt from the turbines.

There are other possible solutions. Smolt go with the flow. If N.B. Power turned off the turbines briefly during the downstream migration in May or June, and water spilled over the dam, the smolt would go safely with the water. There is also evidence that smolt from the Tobique system leave their nursery areas in the fall and begin their downstream migration, spending the winter near the junction of the Tobique and the St. John to shorten their journey in the spring. If N.B. Power were to provide them safe passage, for a brief period every spring and fall the St. John would act like a normal river, except for the delay and confusion that the migrating smolt would experience in the headponds where there is no current to draw them downstream.

"The smolt migration is quite rapid," Ruggles says. "It's feasible to turn the turbines off because it would only be for a short period of time." Another possible solution, he adds, would be to trap smolt above the Tobique Dam and transport them from there to the sea.

"We've probably been considered somewhat negligent by some public groups for not pushing the downstream passage as hard as some think we should have," says Larry Marshall, the federal biologist. "We are making a little progress in that and certainly N.B. Power is being reasonably co-operative. They have been reasonably co-operative in the past on modifications on upstream passage. But from their perspective they can argue that they have already allowed for turbine mortality in the calculations that went into the hatchery they built for us. We haven't always considered ourselves on really firm ground to promote downstream passage because of the fact that presumably it has already been mitigated for."

There are no simple answers on a salmon river. Finding a way to keep the wild smolt alive on their trip downstream will not produce any miracles on the river. Salmon biologist Dr. Alex Bielak says, "It's tragic that we have a river, a great salmon-producing river with a wonderful history, which is in such a grievous state that it's only got 40 per cent of the required spawners, and that's with considerable assistance from the hatchery. I think people have finally realized that this is a resource that we stand to lose. You only have to look farther down south at the $100-million that has been spent to try to bring back the Connecticut and Merrimack and other rivers on the eastern seaboard of the United States to know that once we've lost it, it's a hell of an uphill struggle. I'm still confident we can still turn things around, but we have some fairly significant obstacles in the way."

The three most significant obstacles are at Tobique Narrows, Beechwood and Mactaquac.

CHAPTER FIVE

River Currents

The river people

The fortunes of the Miramichi River's working families rise and fall with the numbers of Atlantic salmon that return from the sea to spawn. This is the story of Keith Wilson, an outfitter whose family has had more than 150 years of history on the river, and Ernest Long, a fishing guide who has a long history at Wilson's: two men whose lives are intimately connected to the Miramichi, and to the Atlantic salmon.

Ernest Long, 62, a renowned fishing guide with a big belly, round face and memorable smile, makes his living from the river. But the river means more to him than money. To borrow the words of Miramichi writer Herb Curtis, the river is Long's sustenance, his entertainment, his theatre.

Keith Wilson, 34, runs the famous Wilson's Sporting Camps. He believes absolutely in the importance of family, history, tradition and the conservation of the resource. He also believes in economic development, and he travels throughout Canada and the United States selling his river. He's full of plans for the future.

The Wilsons are based in McNamee, a small village just upriver from Boiestown. The family's old white farmhouse serves as the camp's dining room, kitchen, office and gift shop. In a glass case on the dining room wall is a homemade book — the history of the Wilsons. Across the road is a cluster of small cabins where the guests sleep. The Wilsons own 16 private pools along a five-kilometre stretch of the Main Southwest Miramichi River.

Ernest Long lives in a bungalow just up the hill. He started doing chores for the Wilsons when he was 13, then began guiding fishermen when he was 16. He joined the army when he was 17 (he lied about his age) and served for more than 20 years, participating in both the Korean War and the Vietnam War (on an international commission with the Canadian army). On every leave he'd end up back at Wilson's, guiding fishermen on the river. When he left the service in March, 1973, he started guiding one month later, on April 15, and since then he has been on the river for 22 straight years. Wilson's is Long's second home.

Ernest Long takes a break beside Home Pool, one of Wilson's Sporting Camps private pools on the Main Southwest Miramichi.

Courtesy of Ernest Long

Long's father was a lumberman. As a boy he thought he would follow his father into the woods, but he fell for the river and has never been able to get away from it. "I can go fishing with somebody and never pick up a fishing rod and I'll enjoy myself as good as the guy who's catching the fish," says Long. "Time don't mean nothing. I can leave home at daylight and come home at dark and wouldn't look at a watch except for you got to be somewhere. It's something you get at and you don't want to quit. I like going to the river. I go to the river every day whether there's fish or not. I was down there this morning looking at the river. I'll be down there again tonight looking at the river."

The river currents have meant not only good times but also tragedy for Long. During a wild spring storm in the 1960s, his grandfather drowned in the Miramichi off the Priceville footbridge, which spans a stretch of water owned by the Wilsons. Five men were on the bridge that night when wild, rising water caught a plank, swung the bridge and broke one of the cables. One man was caught in the bridge wiring. Two fishing guides went out into the turbulent waters in a small boat and cut him free with wire cutters. Another man made it to shore on his own. But the three others drowned — two young men and Long's grandfather, who was in his seventies. One of the young men who drowned was engaged to marry a girl who lived right next to the footbridge. They buried him on the day he was to be married.

*

The history of the Wilsons on the Miramichi began in 1803, when John Wilson arrived from Scotland. In the mid-1800s, Robert Barnwell Roosevelt (an uncle of United States president Theodore Roosevelt) wrote a book entitled *Game Fish of the Northern States of America and British Provinces*. In the book, he described a trip to the Miramichi where he visited the Wilson homestead. "Wilson's habitation was a quaint-looking log house, perched on the edge of a bank overhanging what is called the interval, or fruitful stretch of level land lying between the river and the hills, and its evident antiquity bore testimony that it had belonged to one of the earliest settlers."

Roosevelt complained bitterly about the conditions on the trip and the cost — $306.69 for a month's fishing on the Miramichi, including the round trip from

Boston. He wrote that most of the Miramichi people were shortsighted and overcharged visiting fishermen. "Instead of preserving and increasing the fish, they obstruct the channel entirely with nets, striving by one grand hand to destroy the supply forever. To this general rule Wilson is the only exception, and may be relied on, not only to do whatever in reason is required of him, but to do it at a moderate price."

*

The modern history of the Wilson fishing operation begins with Willard Weston Wilson, who was running a sportfishing camp at McNamee in the 1920s. Willard's wife Sarah did the cooking on an enormous wood stove, which still stands today in the kitchen of the camp. When their oldest daughter, Marie, went to New Hampshire to become a registered nurse, she forged a link between the Wilsons and the United States that is still strong today. After Marie Wilson graduated, she went to work in the infirmary at Dartmouth College in Hanover, New Hampshire. Because she talked about her river, word soon spread about the marvellous Atlantic salmon fishing on the Miramichi. In the following years, Wilson's almost became a private fishing club for Dartmouth people, and the Wilson children started going to Dartmouth College to study. Two of Willard's sons, Willard Woodrow Wilson and James R. Wilson, graduated from Dartmouth in 1941 and 1945.

Three generations of Wilsons: (standing left to right) Bonnie Wilson, Keith Wilson, Ethel Wilson, (seated) Clara Wilson, outside their famous Miramichi fishing camp at McNamee, New Brunswick.
THOMAS R. PERO

In 1948, another of Willard's sons, Murray Wilson, and his wife Clara took over the camp (and hired Ernest Long). In 1968, Murray and his son Karl formed a partnership, with Karl's wife Ethel becoming the hostess; Murray died in 1974. Then, in 1983, Karl Wilson died unexpectedly. The next year, the camp was taken over by Ethel and Karl's youngest son, Keith, who was 22 at the time.

For Keith Wilson, the decision to take over the family business with his mother just seemed like the natural thing to do. The Wilson history says he took over as naturally as a salmon swims upriver to spawn. "My father died and someone had to do it and I just done it," says Keith Wilson; now a father of three children. He lives with his wife, Bonnie, in a new house across the road from the homestead. "It's my heritage. We've been sitting on this property since 1803. This house is 150-something years old, and it really means a lot to me. It's not like I'm a business tycoon of some kind that just moves realty around. It really means something to the family. The stories that I was told at the kitchen table about simple little family events, you just keep repeating them to your children, and hopefully they'll repeat them and they'll have their own stories to tell."

Keith Wilson feels a direct connection to his father, grandfather and great-grandfather. He goes into the basement of the house and sees something his great-grandfather fixed. Or he walks out to the barn and sees some work his grandfather did on the tractor. "Keith couldn't be any more like his grandfather than he is," says Ernest Long. "He's brought that business a long way."

Sportfishermen continue to come to Wilson's year after year, more than 90 per cent of them from the United States. Some of them have left orders in their wills to have their ashes scattered over Wilson's salmon pools. One of the Americans was Arnold Schiffman, who first came to Wilson's in 1972. The 73-year-old Schiffman had made his fortune in the retail jewelry business in North Carolina. A fine fly fisherman, he was ready to try his hand at Atlantic salmon. One summer on the Miramichi hooked him, and he became a regular at Wilson's for the next 22 years. All but three of those years he fished with Ernest Long, and they became close friends. Schiffman usually came up to fish in the spring and fall, and when he wasn't on the river he was often on the telephone, talking to Long about fishing the Miramichi.

One windy day in September, 1992, Ernest Long and Arnold Schiffman (then a spry 93 years old) finally hooked the big one at Big Murphy Pool. When Wilson's guests fish in the fall, they spend the entire day on the river and their guides cook a noon meal on the riverbank. That September day, after Schiffman ate his lunch, he took a nap on the bank, as was his custom. Three other guests were fishing Big Murphy. Long was sitting on the bank watching. Down at the very end of the pool, near a big boulder, he saw a big salmon that kept coming to the surface, not jumping, just rolling its back up out of the water. None of the other fishermen saw it. Long waited quietly for Schiffman to wake up.

"When he came down to the river, I said 'Mr. Schiffman, I saw an awful big salmon out there coming up.' He knows every rock in the pool, too. I said to him, 'Now when we go up in the canoe, I'm going to take you above the big rocks and I want you to get a nice, easy cast going and then just stop and I'm going to drop you down so that when I anchor the boat the fly will hit in the right place."

Long took him out, got the cast going and then he dropped the canoe down. "He threw her over in there, and whammo. So that was it. Everybody just stopped fishing." After a fight of more than an hour, Long netted, then released the monster fish. It weighed 34 and a half pounds. They rested and went out again and "old Mr. Schiffman" hooked another salmon that weighed about 20 pounds. This time, he handed his rod to a friend, who played the fish and lost it.

*

Keith Wilson is one of a generation of young entrepreneurs who will help shape the future of this river. He has purchased the old Canadian National right-of-way that runs along the river past his pools, and he is planning to build a new lodge and a campground for tents and trailers near the river and to rent mountain bikes for trips up and down the river trail.

He is also trying to open up his private waters for New Brunswickers to fish. But they are going to have to pay — $50 a day per rod, and space is limited. "There's a percentage of the population that does resent private ownership," he says. He understands their feeling that their resource is being used by non-New Brunswickers, but he points out that there are economic and conservation considerations. The Wilsons have 16 private pools and fish 12 rods a day. If it were all public water, it would be fished much more heavily than that. He is booked solid for his $370-a-day package that includes fishing, guides, accommodations and meals. He employs cooks, guides and wardens.

"People have to realize that we have to pay to keep this resource," he says. "I was all for it when they raised the fee for trout fishing and then they cut back the limit — they should have done that a long time ago." Long says the money raised should go right back into the river, not into general government revenues. "If we're going to get fish back in the rivers, we can't do it for nothing. It's going to cost a lot of money." Wilson will test whether New Brunswickers are willing to pay a daily fee to fish when he markets his campground. "I want to offer that as a service to New Brunswickers. Naturally I can see an economic opportunity there, but I really, genuinely want to improve relations with New Brunswickers [to repair] that perception that I'm hogging the river and they don't have an opportunity to fish here. I really do want to do something to fix that, so that they can come fishing, that they can feel welcome to come fishing and don't have the perception that it's a rich man's sport. You just can't use something for free without putting anything back into it."

*

Ernest Long keeps meticulous records of the numbers of fish his guests have caught, the size of the fish, the flies they used, and water temperatures. When the fishing gets tough, he refers to his book. Long's log illustrates how the fortunes of the salmon are reflected in the fortunes of the people along this river. In 1982, when the commerical net fishery was pounding the resource, Long guided 68 days and landed 25 fish. "That was almost the end of it." In 1983, he worked 75 days and landed 26 fish. That was the year the river people organized and protested to the federal government to stop the commercial netting. In 1984, the commerical fishery stopped and the killing of large salmon was forbidden. But United States fishermen already weren't returning. Long worked 55 and a half days and landed 52 fish. "From there on it started going up."

In 1985, he worked 69 days and landed 79 fish. In 1986, 77 days, 79 fish. In 1987, 91 days (the fishermen were starting to come back), 149 fish. In 1988, 89 days, 121 fish. In 1989, 97 days, 137 fish. In 1990, 112 days, 108 fish. In 1991, 91 days, 59 fish. In 1992, 77 days, 111 fish. In 1993, 98 days, 131 fish. In 1994, 107 days, 150 fish.

"As you can see, whatever they did was all good. That's all on account of hook-and-release and no commericial salmon fishing. That's the only answer to it." The Miramichi River has come a long way back, but it has by no means come even close to reaching its potential: runs of fish that number in the hundreds of thousands. Ernest Long and Keith Wilson are just waiting for that day.

*

The year after they caught the big one, Schiffman arrived in New Brunswick with a fly rod for Long, a replica of the one he used to catch his big fish. Long stores it in a case. He has never used it and he never will. In July, 1995, I telephoned Schiffman at his home in North Carolina to speak to him about the Miramichi.

"I started out with Wilson and stayed with Wilson," he said. "I liked it. I like the spirit of the place. They're not out to dig you. They charge you enough, yes. But you have fun. I hope I go up there in September. I'm planning to, but I don't know whether my medical director will permit that or not. I'm 96 years old and I know he has questions."

Schiffman didn't make it back. His wife died in August and he died a week later; Atlantic salmon was the second great love of his life. During our conversation in July, Schiffman had kind words for his friend, Ernest Long. "He was born on the river. He knows every hole in it and that makes the difference. I'm very fond of him. He's my buddy. They don't make them any better."

Americans on the river

It is now a week since I returned from my holiday of the year, and a delightful one it has been. I am daily suffering at the thought of Camp Harmony standing vacant on her breezy point, the rippling pool in front whispering its music day after day with no appreciative listener.
— Dean Sage

The clocks in the exclusive old American fishing lodges on the Restigouche River run on United States time. There's an arrogance in that. The American guests start fishing at 10 a.m. New Brunswick time, a bit late to hit the river by our standards. But in the lodge, the clock says it's only 9 a.m. Whose clock do the salmon run on?

In some cases — and Dean Sage's Camp Harmony would be one of these — the Americans have earned a right to a certain amount of arrogance. These Restigouche lodges have a long tradition on the river, a tradition as long as the history of some New Brunswick river families. Their owners have lived and died beside New Brunswick waters. And today the lodges make a major contribution to the province's rural economy, employing more than two people for every one fishing guest. The salmon fishermen put as much as $10-million a year into the local economy in a season that traditionally lasts about eight weeks.

Camp Harmony looks as if it has always graced the bluff that overlooks the confluence of the Upsalquitch and Restigouche rivers. The rivers come together below Camp Harmony's long front veranda in a rush of current and rolling water. From the outside, the old camp looks rustic with its shingled moss-covered roof;

Camp Harmony graces the bluff at the confluence of the Upsalquitch and Restigouche rivers in northern New Brunswick. PHILIP LEE

it was built with wood cut from the mountain behind it and has notched log corners and rough, grey board siding.

The Camp Harmony Angling Club celebrated its 100th anniversary in 1995. The lodge was completed in 1896, the design of renowned New York architect Stanford White, a designer of New York's Grand Central Station, and on the river the creator of the Night Hawk salmon fly. An avid salmon fishermen, White also designed the Kedgwick Lodge, for a member of the Vanderbilt family, and Indian House, one of the Ristigouche Salmon Club's lodges, located near the Million Dollar Pool at the mouth of the Patapédia River. All three White-designed lodges are still being used today.

The Harmony club owns about 6.5 kilometres of private water on the Restigouche and one pool at the mouth of the Upsalquitch. The lodge is built around a comfortable sitting room, which at midday is quiet and feels like a library. Guests stretch out on lounge chairs reading books, talking quietly and drinking iced tea. Fly rods hang outside on wooden peg racks. Some guests sit on the veranda and breath the intoxicating air as it kicks up off the currents of the river.

A July fishing day at Harmony starts this way. At exactly 9:55 a.m. (8:55 a.m. inside the lodge), the camp's eight guides leave their cabin (a building down the hill and separate from the lodge itself) and walk to the dock, where they prepare their canoes for a morning of fishing. Up at the lodge there is a bustle of activity out on the front porch. At 10 a.m. sharp the eight guests, fishing rods in hand, walk down the long stairs to the dock where they wait in a line as the guides bring the canoes to the dock. One by one, each guest steps into a canoe, while the guide holds it steady. Then each canoe motors off, taking one guest and one guide to their own private pool to fish.

New Brunswickers have a big stake in these Restigouche fishing lodges, which have created an atmosphere more luxurious than the Spartan, working-family culture on the Miramichi. But the Restigouche River means everything to many northern New Brunswick families, just as the Miramichi does to many of its families. David Boudreau, manager of Camp Harmony, gestures toward the river and the lodge and says simply: "That's our living." The camp has 17 staff members: eight full-time guides, three wardens, one waitress, one cook, one laundry person, two general chore people and one manager. Harmony's payroll is about $10,000 a week.

*

If Camp Harmony sports a bit of arrogance, the camp's founder, Dean Sage, earned the right. Sage, of Albany, New York — "fierce Dean" as he was known to his friends — was not only a salmon fisherman, but also an author, a boxer and a breeder of trotting horses, dogs and fighting cocks. The son of a wealthy New York merchant, Sage had the leisure time and money to pursue both his sporting and his literary interests.

Sage first came to the Restigouche and Upsalquitch rivers in 1875. He wrote an article about that trip entitled "Ten Days Sport on Salmon Rivers," which was published in the *Atlantic Monthly*. The trip just to get to the Restigouche was an adventure in itself. Sage and his party took a steamer to Boston and then another steamer to Saint John. From the Port City, they boarded a train and rode north to Shediac on the east coast, where they boarded a steamer that travelled up the coast into the Bay of Chaleur to Dalhousie. Then they rented wagons and drove 35 miles to the Matapédia River, where they were met by Micmac river guides. The rest of the trip was made by canoe and horse-drawn barge.

Scenes from the Camp Harmony scrapbook: (top) "Boatmen and their quarters" and "Mickmack Indians." Courtesy of Camp Harmony

On that first trip, Sage arrived at the confluence of the Upsalquitch and Restigouche rivers and described their night there: "We made a raft to transport our luggage down, and that night pitched our tent on a beautiful bluff at the junction of the Upsalquitch and Restigouche, and just over the pool. Here we were comparatively free from flies, with good fishing all about us and a delightful view up and down both rivers." He named the site Harmony after Julia Harmony Twichell, the wife of his good friend the Reverend Joseph Twichell, a clergyman from Hartford, Connecticut.

Sage returned to the Restigouche in 1876 and 1877. These trips were all great adventures, and he started spreading the word about the river. In 1878, he wrote about the area in a letter to his friend David Douglas, a book publisher from Edinburgh, Scotland: "I have, according to the custom of men, been taking a pleasure trip, leaving my wife at home to enjoy the company of the children. My expedition was to the Restigouche River in Her Majesty's late Province of New Brunswick where I passed a fortnight in the best salmon fishing I ever had."

He built his first camp on the Harmony site in 1879, a modest three-room building with birchbark partitions between the rooms. They fished in birchbark canoes with Micmac guides. In 1888, Sage's book, *The Ristigouche and its Salmon Fishing,* was published in Edinburgh by David Douglas. Seven years later, the Camp Harmony Angling Club was formed, and in 1896, the majestic new camp was completed. Women were allowed at the camp every summer, but only after July 1. Sage's wife, Susan, was a regular guest at Harmony along with some of their children; in 1884, their eldest daughter brought along a banjo, which (Sage later wrote) had to be "suppressed."

Micmac guides ran the camp in its early years, but at some point they left. Local legend has it that one of the guides took a swim in the spring water tank uphill from the lodge and drowned. No one knew where he was and the guests drank the water for a week. That, apparently, was the end of the Micmac presence at Harmony.

*

Camp Harmony Angling Club allows only seven members. Today, two are from Texas, one is from Philadelphia, one is from Delaware, two are from Detroit and one is from New York. Only club chairman Bill Flemming, a Texas lawyer, has a direct connection to the Sage family. A new member, who must receive the unanimous approval of the group, pays $25,000 (U.S.) to join the club. Then, members pay big money to fish at Harmony — about $6,600 (U.S.) for a couple for a week. This price includes everything: guides, food, liquor, salmon flies. Only eight guests are allowed at a time. The operation of the camp during the 1994 season cost more than $215,000; the grocery bill alone was more than $17,000, and all supplies were purchased locally. The property tax on the more than 400 acres of land and private pools is about $10,000 a year.

David Boudreau, 43, has been working at Harmony since 1979. The camp has had good fishing years and bad. "It's been going up and down for the last while," he says. There used to be an early run of fish on the Restigouche, big fish that had arrived by May 24. These were 40- to 45-pound salmon, good spawners for the river. "There's no more of them," and the grilse population is also dropping. Poor fishing years make it hard to book the club.

Boudreau says there is little local resentment of operations like Camp Harmony. "The local people all know people need jobs." And to keep public relations healthy, in August when American guests no longer fish the Restigouche, Boudreau gives local people free permits to fish the private waters.

*

Dean Sage returned to Camp Harmony every summer. On the morning of June 23, 1902, Sage and his guide Alex Marchand fished the river and landed three salmon. Sage returned to Camp Harmony in good spirits and ate lunch. About 4 p.m. he began to complain of chest pain, and soon after that he died of a heart attack as he lay in his room at the camp. According to Sage's friend Twichell, during the night a casket was brought up from Matapédia, about seven miles downstream, upon two canoes fastened together, and he was "borne down the stream over the flowing waters between the leafy banks so familiar to him." An impressive scene to contemplate, he added. A sad journey, but not without beauty.

RSC

At high noon beneath the sultry summer sun at the oldest and most exclusive fishing club in North America, the guests lounge on the veranda, drinks in hand, waiting to be called to lunch by waitresses in white uniforms. The long, white lodge with green trim, surrounded by a lush garden and bordered by a fence, is situated just behind the train station in Matapédia, Quebec. A short walk downhill from the train station is the Restigouche River and a row of river canoes on a rocky beach.

A modest sign in front of the property says simply "RSC." But this is no modest club. This is the Ristigouche Salmon Club. I was in Matapédia to interview Al Carter, the club's manager for the past 19 years and a man who knows as much about the Restigouche and salmon as anyone on the river. Al Carter (the brother of Dr. Wilfred Carter) works out of an old white house on the club grounds.

When I arrived, Carter was taking the children of his guests out to lunch at McDonald's across the Quebec-New Brunswick border in Atholville — one of his yearly rituals. When he returned, we sat down in his office and he told me of his deep concerns for the future of the Restigouche River and its salmon runs. The club has records of every fish caught on club water since 1880. In its own record books, the club can see the resource disappearing. Carter also says the river is becoming crowded with unregulated canoeists, who are leaving fields of garbage

The Ristigouche Salmon Club's Pine Island Lodge. RSC

behind and interfering with salmon anglers. He has expressed his worries about the river repeatedly to the New Brunswick government, but he says the politicians just don't seem interested in hearing what he has to say.

"We know the fish stocks are in decline," he says. "We take the fish count every day. We can see the trends. It gets frustrating at times to see all this happen, to talk about it. We have had everybody on the river, from [New Brunswick premier] Frank McKenna to every natural resources minister who was ever appointed, and we tell them the same thing. It doesn't seem to make much difference. It goes in one ear and out the other. They keep giving us excuses, such as, 'We can't put regulations in until we can afford to enforce them.' We say, 'You're going to have nothing to enforce if you don't do something.'"

Carter says he is concerned about ocean nets that are killing salmon at sea and native nets that are catching fish at the mouth of the river. Some of these threats may be beyond our control, he says, but when the salmon hit the river, they are ours to protect. "From the mouth of the river up, there's no one else who controls them except the citizens of New Brunswick. And we don't do that very well. We're killing it."

*

Peter C. Newman, in his book *The Canadian Establishment*, lists the Ristigouche Salmon Club as one of the two most exclusive "sportsmen's hideaways" in Canada, clubs "that lure the international business celebrities." The Ristigouche Salmon Club, the oldest registered fishing club in North America, has been luring celebrities since it was formed in 1880. It was started by a group of Americans who came up from Boston and New York in sailing ships and anchored them in the Restigouche off Campbellton. They dropped little boats off the sides of their ships and went fishing for Atlantic salmon. In the early days, they didn't have to travel too far to fish. But as they went farther upstream, they pitched tents and built camps. They also started buying up land.

Eventually, the 38 fishermen who had establishments on the river realized that they all had similar needs. Al Carter says: "Because they all had the same problems, they all got together and said, 'Let's centralize this and put ourselves into a group and maybe somebody can get ice for us all, somebody can get firewood for us all.' And they formed a club. They put all of their assets together and named it the Ristigouche Salmon Club. Those people had purchased tracts of land all the way along the river and those all fell into a big pot. We own property on both sides of the river for miles. They come with riparian fishing rights. You wonder about the value of those fishing rights today when you look around."

When sailing ships were no longer used, the club members left their private railway cars at the station in Matapédia, a stone's throw away from the main lodge. These days, members sometimes come in private jets in the morning, go fishing and fly home at night. Today's membership includes the elite of the Canadian and U.S. business establishment. Over the years, the club has hosted many famous people, including Bing Crosby, Prince Charles, former U.S. presidents Herbert Hoover and Jimmy Carter, golf legend Jack Nicklaus (who is a regular visitor), U.S. General Norman Schwartzkopf and a series of Canadian prime ministers and premiers, not to mention a regular flow of the most important business executives on the continent.

The RSC now has 30 members, half from Canada and half from the United States, and employs between 40 and 45 people for a 12-week season. Its guides are the highest paid on the river system. Some employees work for the club all winter, cutting wood and preparing ice.

Members of the club and their guests fish along about 64 kilometres of river. The management of the club is careful to preserve the natural state of the river banks. If a tree falls, workers leave it there. Firewood comes from tree stands over the mountain, far from the river. "It's worth a person's job to cut a tree around here," Carter says.

❋

The Ristigouche Salmon Club and other riparian owners operate and finance their own river protection system: the Restigouche Riparian Association. The association has 52 staff members, including one bookkeeper, an executive director, two warden supervisors — one for the upper river system and one for the lower — and 48 uniformed wardens, more in this watershed than the federal and provincial governments combined.

For the past 18 years, Bill Murray has been running this non-profit association out of his home in Campbellton. Day and night, the association guards 80 kilometres of water on this wide, clear, mountain river, which rises in northern New Brunswick from the Kedgwick Lakes and brooks in the Saint-Basile area and empties into the Bay of Chaleur. "The river is a wonder of nature," says the soft-spoken Murray, the association's executive director. "All we have to do is look after it a little bit and it will look after itself." The association has 23 members, all of whom own or lease water in the Restigouche watershed.

The association was founded in the late 1800s when some river owners banded together to combat poaching and net fishing by forestry companies that were looking for a cheap way to feed their lumberjacks. By 1895, it was a loose organization, and in 1912 the association was incorporated and received a federal charter as a non-profit organization.

Clouds reflect on the surface of the Restigouche River. DAVID NICKERSON

In 1915, the association hired 59 wardens who worked to end poaching, especially the dynamiting of salmon pools. In its early years, the association also bought out the traditional salmon-netting rights in the river upstream from Campbellton. Today, the riparian association has an advanced communications system, with transmitters on Squaw Cap and Kedgwick Mountains, that keep Murray in close contact with his wardens, strategically scattered throughout the watershed.

The most carefully guarded pool on the river has always been the pool at the mouth of the Patapédia River owned by the Ristigouche Salmon Club. The pool was purchased by the club in the late 1800s for $250. Today, it is known as the Million Dollar Pool, and it sometimes holds thousands of salmon at one time in its depths, cooled by the waters from the Patapédia. "How much is the Million Dollar Pool worth?" Murray asks. "A hell of a lot more than a million dollars."

*

I met Al Carter on a Sunday. Early that morning he had travelled down to Matapédia from upriver and had inspected the river after a nice sunny summer weekend — it wasn't a pretty sight. "If you could see the campground there today you wouldn't believe it. There must have been 200 tents there last night, and it is one big litter ground today. The bags have been just left there and the birds have picked the bags apart and the garbage has been scattered all over. We will clean that up tomorrow. We cleaned it up last week and it took two men three days."

The Ristigouche Salmon Club's Indian House Lodge, a short canoe ride from the Million Dollar Pool. RSC

The club maintains the campground at Cross Point as a condition of its lease of salmon water from the province, an agreement worth more than $500,000 over 10 years. Carter says he has no objection to keeping up the campground. "We bid [on the lease] with our eyes wide open. We supply picnic tables, toilets, fireplaces and firewood. There's a regulation for us that says we have to provide a campground. But there's no regulation for the people that says you have to go there. And there should be. What the hell's the use of telling us we have to build a campground if you can go down and camp across the river from it? The canoes are not regulated at all. And we have said for years that unless the canoes are regulated, we cannot fish."

On most days during the summer, hundreds of canoes are floating down the Restigouche. On long weekends, the river is crowded with canoes. New Brunswick's Natural Resources Minister Alan Graham says the province is promoting the river as a canoe destination, and there will be more canoes in the future. He agrees that canoes have been a major problem on the river and that "we've got to clean up our act."

A stretch of the Restigouche River has been nominated to become a Canadian Heritage River, and this raised its profile further. The province has outlawed drinking alcohol in canoes and has enlisted the help of the RCMP to enforce the rule. "There has been a real, marked improvement," Graham says. "We've come a long way, but I'm not satisfied yet. It is a beautiful river and it has to be maintained." Graham doesn't see the need now for any river regulations. He says his officials are trying to "educate" people about how to treat the river before they put their canoes into the water.

Al Carter says 95 per cent of the people who use the river are not a problem. The club puts markers in the salmon pools with arrows showing canoeists the channels. Most people shut off their motors and float through a pool when the club's guests are fishing. But a small, loud minority go though the pools with their motors at full throttle, bump the fishermen's boats, throw beer bottles in their canoes and shout at them. "Everybody who is a non-New Brunswick resident is a damn American. It's all, 'Go home, damn Americans. Leave the river to us. It's our river.'"

⁂

Al Carter says the attitude of some New Brunswickers toward the river can be seen clearly in the story of Red Bank. The club used to have a Crown lease called Red Bank, located about two miles upstream from the Patapédia, where it built one of its classic old lodges. Some years later, the province decided to turn Red Bank into a Crown Reserve stretch, special salmon water leased daily to New Brunswick fishermen through a lottery system.

According to Carter, the club was disappointed but took the decision in stride. The lodge was no longer of any use to the club, so the staff removed the club's possessions and offered to sell the building to the province for a small fee. The province didn't want to buy it, but the club said to go ahead and use it anyway. The fishermen who were fishing at Red Bank started to use the camp. "And then they started to use the doors and windows, balcony and steps to make bonfires. And they tore off all the plumbing fixtures. People went in and took all the plumbing and all the piping and all the electrical fixtures and all the windows. Within three months it was nothing but a shell and it's now on the ground. And it's a shame." These days, fishermen at Red Bank sleep in tents.

⁂

The Ristigouche Salmon Club and other clubs along the river have tried to buy out the native gill-net fishery at the mouth of the Restigouche. Carter says the clubs on the river were prepared to pay the natives market value for every salmon they caught and released in trap nets. However, the Micmac have insisted on exercising their right to gill-net wild Restigouche salmon and kill them. "We're trying to convince them in every way we can that the female salmon need to be saved. If you want to save the river, you have to save the breeding stock."

Carter is also worried about high-tech ocean fishing that he thinks is killing salmon and their food supply. He says the native and offshore salmon fisheries make it difficult to justify stocking programs like satellite rearing, in which young salmon are raised on the banks of the river in tanks. "We take thousands of small fish and look after them and feed them and empty them back into the river in the belief that a large number of them are going to come back into the river that they were spawned in. However, the majority of them are going to get caught before they have a chance to enter back into the river. And that's pretty discouraging. I hate to see it go. I've fished salmon all my life. I'm glad I've got 50 years of fishing behind me, not ahead of me. Many things concern us. And all we're trying to do is have a little fun and go fishing."

Larry's Gulch

The New Brunswick government's own exclusive lodge is nestled on a high bluff above the Restigouche River, a short trip along winding gravel roads from the town of Kedgwick. From the front lawn of Larry's Gulch you can look down into the clear waters of the river. The lodge is a cluster of well-maintained, brown-shingled buildings and cabins. The main lodge includes a dining area, a sitting room and a bar with a stone fireplace. There is an icehouse at the rear of the buildings to preserve the salmon killed by the guests.

Although the lodge holds 16 guests, only eight rods are allowed to fish for salmon at one time, and eight canoes with motors are beached below the lodge. Larry's Gulch water includes eight major salmon pools — private fishing for the guests of the New Brunswick government. From 1990 to 1995, 67 per cent of the guests at the lodge were treated to their fishing trips by taxpayers — at a cost of about $800,000. The rest of the anglers, most at Larry's Gulch by special invitation, paid their own way: $450 a day, which includes fishing, canoes, guides, meals and accommodations. Fishing guests bring their own liquor. The provincial government has owned and operated this lodge for more than 20 years. The assessed value of the camp today, including water rights, is $951,000, but it would sell on the open market for a lot more than that.

In response to a Right to Information request seeking details about the operation of Larry's Gulch, Natural Resources Minister Alan Graham wrote that the lodge operates "primarily to ensure access to a facility that can be used to entertain guests who may potentially invest in the Province of New Brunswick or may provide future opportunities to the Province."

Every year, the ministers of various provincial government departments book their space at the lodge. Then the remaining fishing spots are reserved, most often by friends of the government who are looking for a chance to fish there. Cabinet ministers submit a list of names for bookings every year. Larry's Gulch is almost always full, and it doesn't have to advertise.

*

Larry's Gulch Lodge became provincial government property in 1974; the waters themselves have been in the province's hands for many years. Dr. George Frederick Clarke writes in *Six Salmon Rivers and Another* that in 1918 the provincial government decided to create some public fishing water on the Restigouche, which in those days was almost entirely leased out to private interests. The public water was first operated by an outfitter named Jack Russell, who charged rod fees. Anglers who fished the water tented on the river bank until Russell "built a

From the lawn in front of Larry's Gulch Lodge a visitor can look down into the waters of the Restigouche River. PHILIP LEE

wooden camp on a high knoll on the opposite side of the river, into which water was conveyed from a nearby spring brook," Clarke writes.

The four miles of public river was called the Government Open Water, and during Clarke's day it was operated by outfitters David and John (Jock) Ogilvy. (The famous Chestnut Ogilvy river canoe is named after that family.) Clarke fished these waters without a break for 30 years. He writes that, in 1952, the Government Open Water was taken over by Alvin Murray, a renowned river guide who was manager of the Restigouche Riparian Association for many years. (His son, Bill Murray, is now the association's executive director.) A few years later, Murray obtained the Ristigouche Salmon Club's beautiful lodge just below Downs Gulch Brook.

In 1974, Premier Richard Hatfield bought Murray's lodge for $40,500, renamed it Larry's Gulch (the original name of the stretch of water), and formed a private government fishing club. The *Telegraph Journal* reported the purchase: "VIP fishermen from Canada had the opportunity this summer to fish the waters that until this year were the exclusive preserve of American millionaires. The New Brunswick government earlier this year bought Murray's Camp and renamed it Larry's Gulch, and is now using it as a fishing camp for provincial government senior officials, MLAs, cabinet ministers and guests invited to the pool."

Northern New Brunswickers complained that there was no more public water on the Restigouche. They had a good point. There were few Crown Reserve stretches of salmon water in the early 1970s, and the sort of arrangement that allowed the public to fish for daily fees had disappeared, replaced in most cases by exclusive private fishing clubs.

In the summer of 1974, soon after the purchase of the lodge was announced, a protest group descended on Larry's Gulch. Some protesters fished and hooked salmon, but just what happened on the day that became known as the Battle of Larry's Gulch is disputed to this day. At the centre of the dispute is the version related by veteran journalist Charles Lynch in his book, *Fishing with Simon.* (The Simon in the book's title is Simon Riesman, the former deputy minister of finance in Ottawa and later the country's chief negotiator of the free-trade agreement with the United States.)

An angler works one of the New Brunswick government's pools on the Larry's Gulch stretch of the Restigouche River. DEPARTMENT OF NATURAL RESOURCES

Lynch writes that he visited Larry's Gulch the year Hatfield's government bought it. "Almost all the best salmon water was in the hands of rich big shots, most of them Americans, who bid for the leases at the auctions and managed to keep the best waters to themselves. It was Premier Richard Hatfield's inspiration to buy one of the exclusive fishing camps on the best New Brunswick river, the Restigouche, and so it was that Larry's Gulch came to be provincial property, the government taking over the camp buildings, the staff and six salmon pools on a two-mile stretch of the river. The acquisition was not widely publicized at the time, nor has it much been written about since."

According to Lynch, Riesman was among the first group of fishermen to visit the camp, and he invited Lynch to come along. Riesman was invited because Hatfield was looking for a grant to bail out the failing Bricklin sports car industry — a boondoggle that had cost provincial taxpayers $30-million. When Lynch and Riesman arrived at the road to Larry's Gulch, they found a long line of vehicles parked along the shoulder. "When we got close to the camp, there were police cars and people shouting, and much milling about. There were men, women, and children, and a proliferation of fishing gear that didn't go with the image of the Restigouche as a pristine river where everything but fly-fishing was forbidden and even that was closely regulated." Lynch maintains that there were about 200 people there, on the shore and on the river in boats. Lynch writes:

> They had read a paragraph in the paper, saying that Larry's Gulch had been bought by the province, and had decided that if it belonged to the government, it belonged to them. And by God, they were going to fish it, a privilege denied not only to them but to their fathers and grandfathers. Having poached the river in the dark of night, they were going to fish it in the open. The townspeople of nearby Saint-Quentin,

> Kedgwick and points east and west unlimbered their gear and fished through the day. In the process, they showed the salmon an assortment of hardware not seen in those sanctified waters before or since — spinners, minnows, worms, and maybe even a fly or two, though fly-fishing is not a common method among the rural folk of northern New Brunswick, especially when they are fishing for the pot. Nobody will ever know how many fish were taken that day by the multitude, but it is presumed there were enough to go around and to feed everybody who came, with some left over for sale, as is the custom.

After an excerpt from Lynch's book appeared in the *Atlantic Salmon Journal*, an angry Paul Bourgoin, president of the New Brunswick Wildlife Federation, responded with a letter that called Lynch's piece a fairy tale. Bourgoin said Larry's Gulch was purchased on March 12, 1974 and that the Battle of Larry's Gulch took place on a Saturday in July, 1974. He added that Riesman was a guest at Larry's Gulch in the summer of 1975 and therefore couldn't have been present at the Battle.

According to Bourgoin, there was a peaceful demonstration that day — 53 people protested and three people fished on the government's new private waters. What resulted from this demonstration, he said, was the creation of some open water on the Little Main Restigouche in 1975. "To go as far as labelling the rural folk of Northern New Brunswick as poachers — well, I find this very distasteful and downright racist on the part of the author, who does not appear to have the faintest idea of when and what happened that day." One salmon was hooked and released, he said, and for Lynch to say everyone took a salmon home is "bigotry."

*

Quebec writer Mordecai Richler, who is a passionate Atlantic salmon fisherman, gives us a much more enlightening account of a trip to Larry's Gulch in his novel *Solomon Gursky Was Here*. Richler writes of the drive from Quebec to the Restigouche and "Vince's Gulch" — a thinly disguised Larry's Gulch.

> Rising and dipping he spun into the valley of the Matapédia, river banks soaring like canyon walls, the spruce and cedar and birch not so much rooted there as scaling the cliffs on which they held no more than a tenuous toe-hold. Then he crossed into New Brunswick at Pointe-à-la-Croix, taking the bridge into Campbellton and then making straight for the camp of the Restigouche. Vince's Gulch was made

> up of a dining lodge and a sleeping lodge and a spread of outbuildings, including an ice-house.

Richler describes the first night of fishing experienced by his character Moses:

> A half hour passed and then the deer flies came out and it began to drizzle. Covering the far fast water, stripping his line quickly, Moses got his strike. A big fish, maybe 30 pounds, taking so hard Moses didn't even have to set the hook, his rod already bent double. Immediately the line screeched and the fish shot downriver, taking most of Moses's backing before it paused and he started to reel in the slack.
>
> Jim lifted the anchor and began to paddle gently toward shore, his net within reach. The fish came close enough to look at the canoe and raced downriver again, breaking water about fifty feet out. Flipping in the air. Dancing on its tail.
>
> "Hey there, Moses. Hey there."
>
> The fish struck for the bottom and Moses imagined it down there, outraged, rubbing its throbbing jaw against the gravel, trying to dislodge the hook. It couldn't, obviously, so it gave in to bad temper, flying out of the boiling water once more, shaking its angry head, diving, then resting deep, maybe pondering tactics.

Moses lost the fish. When he returned to the lodge, he witnessed what the McKenna government today claims Larry's Gulch is really all about. "Entering the dining lodge in search of a coffee, Moses found that Barney and Larry were being entertained by a deputy of New Brunswick's minister of trade, an obsequious young man wearing a tartan jacket and canary yellow Bermuda shorts. The deputy had come equipped with information on local land and labor costs. Larry, taking notes on a legal pad, needed to know what kind of sweetener they could expect investment- and tax-wise from the provincial government."

CHAPTER SIX

First Nations

Salmon self-government

We the indigenous people of Gespe'gewa'gi are vested by Gisiteget with sacred responsibilities for stewardship of the land, waters and all living things.
— Preamble, Listuguj Mi'gmaq First Nation fishing law.

The leaders of the Listuguj Mi'gmaq First Nation say their Atlantic salmon gill-net fishery in the Restigouche River estuary is a pure expression of native self-government. There are about 2,600 Micmac in this band, whose reserve is located at Pointe-à-la-Croix on the Quebec bank of the Restigouche River, directly opposite the city of Campbellton, New Brunswick. In 1992, the band declared that it was responsible for its own salmon fishery — and finally there was peace on the river.

The native salmon wars that have raged for more than a decade on Atlantic Canada's rivers, and nowhere more fiercely than at Pointe-à-la-Croix, have cost Canadian taxpayers millions of dollars, dominated the time of our provincial and federal conservation officers and tied up our courts with hundreds of illegal fishing cases. The salmon wars have always had more to do with native rights than with fishing and food. Unfortunately, this power struggle has been played out with a threatened resource, and the Atlantic salmon may end up being the big loser.

Today, the Micmac of the Listuguj First Nation say they have found a way to hold on to their right to fish, to create peace on the river and to conserve the salmon. Their critics say their plan has a major flaw — their deadly gill nets are still killing large salmon on the river. But the Micmac insist their fishery is being carefully controlled and is not harming the resource. This first nation is run by a group of focused, intelligent and articulate leaders who cannot be ignored.

The philosophy behind the band's approach to managing the salmon fishery is laid out in its fishing law, which opens with an affirmation of the "sacred responsibilities" for stewardship of the river that Gisiteget, the creator, has vested in the people of Gespe'gewa'gi, the last land. The days of an unregulated, free-for-all fishery are over, native leaders say. The band's fishing law regulates where and when the gill nets can be in the river, but it sets no quota for the fishery. "It's an expression of self-government," says Beaver Paul, a former band policy adviser who now works for the band's economic development corporation. "I've never seen a clearer or more definitive expression of self-government than that document. We know that we're on the cutting edge. We know that we're on soft ground, but that doesn't frighten us. We believe we're doing the right thing."

The 1992 Listuguj fishing law states that, although the ancient treaties between the British government and the Micmac people give them the right to fish salmon throughout the Restigouche watershed, they will restrict themselves to a carefully managed fishery along 15 kilometres of water near the mouth of the river. "We're willing to share that law with any other first nation," Paul says. "But you have to be at a certain stage to understand what that's saying. If you're just beginning your journey, that may be going too fast."

*

The journey of the Listuguj Micmac has been filled with conflict. In the past, the Listuguj First Nation signed a number of salmon-fishing agreements with Ottawa, deals that involved strict quotas on Micmac salmon catches and as much as $500,000 a year in government grants to create jobs for their people. Largely because of the inability of the two sides to agree on anything, the Restigouche band and Ottawa were often at war. In June, 1980, the Micmac blocked the northbound lane on the bridge that connects Campbellton and Quebec to protest what they felt was harassment in their fishery: Government officials were searching vehicles for untagged salmon that were being shipped out of the reserve to be sold. A mob of Campbellton residents responded to the blockade with baseball bats and tire irons and 11 people were arrested. Then came the infamous raids of 1981.

In June, 1981, Quebec government officials raided the Restigouche salmon fishery with boats, police cars and helicopters and seized more than 140 nets. Gary Metallic, a salmon fisherman and band councillor who is one of the architects of the new, independent Listuguj fishery, went to jail in 1981. For three days before the raid, government officials demanded that the native fishermen remove their nets from the water. Metallic and the other fishermen refused. Metallic said that all he felt in 1981 was a deep anger toward these outsiders. "It was like a military type of manoeuvre," he says.

In 1986, the band agreed to a quota of 1,184 salmon, with the stipulation that they be tagged and registered on shore. However, provincial government observers with high-powered scopes staked out the opposite bank and saw many salmon being killed but not tagged. The next year, natives pelted a government helicopter with rocks when it tried to arrest two Micmac for illegal fishing. From

1989 to 1992, 130 Listiguj Micmac fishermen were charged by federal officials with illegal fishing. About 40 of these cases were still working their way through the courts in 1995. "We realized that something was really wrong here," Metallic says. In the summer of 1995 on the Northwest Miramichi, New Brunswickers witnessed a prolonged war at Big Hole Tract between Ottawa and Micmac who said the federal government was interfering with their right to fish. The Micmac at Listuguj have great sympathy for the natives who staged the armed standoff on the Miramichi. They have sympathy because they've been there themselves.

As for the millions of dollars paid to the Listuguj reserve during the time it was signing fishing agreements with Ottawa, the band really has nothing to show for it, Metallic points out. What happened to the money? "That's the million-dollar question," he answers. A new chief and council was elected in 1992. These new leaders soon informed Ottawa and the provinces that they would sign no more salmon-fishing agreements. The band promised to fish in a way that would conserve the salmon.

*

When it declared salmon self-government, the band was on solid legal ground, thanks to an aboriginal British Columbian named Ronald Edward Sparrow, a member of the Musqueam Band. When Sparrow strung his drift net in British Columbia's Fraser River in May of 1984, he provoked a legal battle that would result in a crucial Supreme Court of Canada decision on native fishing rights. The facts of the case are simple: Sparrow was arrested in an area called Canoe Passage in the Fraser River system and charged with fishing with a drift net longer than the nets permitted by the band's Indian Food Fishing Licence. Six years later, on May 31, 1990, Sparrow won his case.

The Supreme Court did not decide whether Sparrow was guilty or not guilty of illegal fishing; it said it didn't have enough information to make that decision. The court did set out a framework to guide Canadians on the subject of native fisheries. Native fishing rights are grounded in aboriginal rights and treaty rights, both of which were confirmed when the Canadian Constitution was patriated in 1982. Aboriginal rights confirm that natives, as the first people to occupy this land, have the right to use the fruits of the land, including fish. Treaty rights are based on long-standing agreements between individual native tribes and the British or Canadian governments. Some treaties include fishing rights.

In the Sparrow case, the Supreme Court determined that Sparrow's band had a constitutionally protected aboriginal fishing right. The decision drew the following conclusions for all native fisheries from that basic fact. Where you can establish that a traditional fishing right existed in 1982 (that it hadn't in any way been

The Listuguj Mi'gmaq First Nation operates trap nets like this one for tagging salmon on the Restigouche River. Band members kill salmon in gill nets farther downstream.
PHILIP LEE

abolished or traded away over the years), that right is protected from unwarranted interference by government. The traditional fishery is a community right, not an individual right. Government can regulate the native traditional fishery to conserve fish stocks; thus, traditional fishing rights can be confined or suspended, but never abolished, and the government has to demonstrate a good reason to limit fishing.

"One clear reason that would justify such a suspension or regulation or limitation is the interest of conservation," says Gordon Gregory, a Fredericton lawyer and expert on native fishing rights. Gregory has provided advice to the Atlantic Salmon Federation and other clients on the issue of native rights, and he was intimately involved in native fishery issues during his tenure as New Brunswick's deputy minister of justice.

The Supreme Court said it did not have enough facts to determine whether the limit on the length of drift nets like Sparrow's was driven by conservation, so it sent his case back to trial, asking the federal government to prove its case. The charges against Sparrow were dropped.

The court recognized that there are competing interests among commercial, sport and native fishermen, and it stated that the order of priorities is conservation first, and the native food and ceremonial fishery second. "All other potential users rank after that," Gregory explains. "So if there are in a river enough fish only to meet conservation needs and to maintain a species, then [the government] can close the river to all fishing. If on the other hand there is a surplus in a river, but only a modest surplus sufficient to provide some food fishery for the Indians in question, then the Indians come first and no one else can take. The court ruled that native fishermen don't have to use traditional fishing methods. They are entitled to use whatever modern means are available to exercise that right."

The court stated that, before setting regulations, the government is expected to hold discussions, negotiations and consultations with the natives. The government is not to act abruptly and arbitrarily. Gregory points out that since the Sparrow decision, the burden of proof first rests with the native and then is passed on to the Crown — the native demonstrates his right, then the government demonstrates the need to regulate that right.

Thomas Latter (left), a fish technician, and Wendell Metallic, a band councillor, return from one of the Listuguj Mi'gmaq First Nation trap nets on the Restigouche River. PHILIP LEE

In 1992, the Department of Fisheries and Oceans came forward with an Aboriginal Fisheries Strategy that followed the guidelines set down in the Sparrow decision. The department said it would give native bands communal licences to fish. The agreements, negotiated each year, would establish fishing limits and would also involve conservation, fish counting and habitat restoration.

*

Armed with the legal backing of Sparrow, the Listuguj band ignored Ottawa's bureaucrats and took control of its own fishery. "The community was going to take over management of a certain identified area of fishery," Metallic says. "We went to the fishermen, we went to community members, and told them that, in the area of conservation, responsibility has to be first taken by us because if we didn't we would be opening the door for either the provincial or federal organizations to come in and do it for us, as they did in the past."

The Listuguj fishing law addresses conservation, management and protection, and applies to the entire Restigouche watershed, including the Restigouche, Kemp, Patapédia, Matapédia, Upsalquitch and Kedgwick rivers and all their tributaries. The first nation government is entirely responsible for the management of the fishing plan, but it may enter into management arrangements with other governments "in the interest of conservation and management of fisheries resources."

The band's Natural Resources Division is responsible for stock assessments, harvest limits, special protection zones and river enhancement. The new Mi'gmaq Rangers are responsible for monitoring the river, providing surveillance, conducting investigations and developing education programs. The uniformed Rangers have high-tech speedboats as well as canoes to patrol the river, and they are based in a new building on the reserve that also houses a native police force, fire department and ambulance service.

The Rangers also protect the native fishermen from outsiders. The Listuguj fishing law states: "No person other than a duly appointed Mi'gmaq Ranger shall interfere with or molest the Mi'gmaq in fishing and conducting fishing-related activities, including landing, storage, transport, use and disposal of fish by sale or otherwise." In 1995, the Quebec government paid the Rangers $400,000 to police the river.

The native fishermen have no quotas. No limits will be imposed unless the band's own Natural Resources Division decides that the native harvesting activities "are exclusively responsible for endangering fish stocks." Any Micmac who is not a resident of Listuguj may harvest salmon in the Restigouche after giving notice to the first nation government.

In the estuary, native fishermen can catch salmon with rods and reels, spears, trap nets and gill nets. Gill nets, which are restricted to a maximum length of 500 feet, can remain in the river for five days, after which they must be pulled out of the water for two days. Gill-netters must maintain log books to record their catch. There is no netting allowed from 8 a.m. Mondays until 4 p.m. Wednesdays, and on the other days of the week fishing is allowed only between 4 p.m. and 8 a.m. If a fisherman violates these rules, the penalties are simple. For the first offence his net is taken away and returned after two days. For the second offence, his net is taken away for a week. For the third offence, the net is taken away until the season is closed and then it is returned.

There is also a limited area in which to place the nets. For example, nets are not allowed in an area of the river called "the bottleneck," where thousands of salmon run through a narrow river channel. The Rangers patrol that area carefully. The band fishes one side of the river only, although the Micmacs say they clearly have the right to fish both sides.

The band is sensitive to questions about the numbers of fish being killed in the gill nets. "Why do you ask about numbers?" Paul asked during an interview on the reserve. "Do you ask that out of the interest of Canadian society? Those type of things have to be put into context — the size of the stock, the type of harvesting that goes on here, the type of harvesting that goes on on the river. They take 1,400 to 1,700 salmon out of the Matapédia. That's not what we take down here. We take about 1,200 to 1,400."

Beaver Paul says the band plans to operate a sport-fishing camp upriver; they have the money and are just waiting for the right opportunity. "That's going to be a business and it will be run like a business. We'll buy it. We'll act just like anybody else. Once we're off-reserve, it's a business."

According to Gary Metallic, the Listuguj First Nation is working to be accountable both to natives and non-natives. Until natives clean up their act politically, self-government will be impossible, he argues. "We are showing them that we can be accountable and we're willing to put everything in place that Canada and the provinces have." This includes law and order, community management and salmon conservation. "If you don't do that your credibility doesn't exist. And we are tired of that. We are tired of being portrayed as the unreasonable ones. We are trying to present this as a model community."

When Rights Collide

When David and Rhonda Pelletier fished for Atlantic salmon in the Restigouche River in July, 1993, they had no idea that they were casting doubt on private property rights on New Brunswick's rivers.

About 30 metres downriver from the mouth of the Little Main Restigouche, the Pelletiers parked their Chevrolet truck and walked to the river past bilingual No Trespassing signs. Game wardens saw them fishing for salmon with spinning gear — an illegal form of fishing on salmon rivers in New Brunswick. Spinning for salmon, however, is not illegal for everyone. The Pelletiers, brother and sister, are Maliseet Indians from St. Basile; natives in New Brunswick, because of their treaty and aboriginal rights, can fish for salmon on Crown-owned waters with any type of fishing gear they like. But the Pelletiers were fishing on riparian waters, the private property of the Kedgwick Lodge.

The province decided to take them to court on charges of illegal fishing, arguing that the Crown land grant, now owned by the lodge, which included exclusive fishing rights, extinguished native treaty rights. Therefore, the province argued, the Pelletiers were fishing illegally because on private property they were subject to the same rules as everybody else in New Brunswick. An argument with wide-ranging implications for private property rights, riparian rights, and treaty rights in this province was going to be tested on an obscure — some might say trivial — case of illegal fishing.

The Pelletier case went to trial on September 13, 1994, in Campbellton, with Keith McCormick arguing for the Crown and Richard Hatchette for the Pelletiers. When Hatchette presented his closing argument, he made a statement that would send chills up the spines of riparian owners all over the province: "I will try to be brief, but there are so many defences in this case that it boggles my mind."

On February 9, 1995, Judge Pierre W. Arseneault found the Pelletiers not guilty of illegal fishing, and he rejected the Crown's argument that the treaty rights had been extinguished on private property. "There is not sufficient evidence to conclude, beyond a reasonable doubt, that the treaty has been extinguished at the location in question," he said. The decision stunned riparian owners.

"The provincial government insisted on forging ahead on this case," says Bill Murray of the Restigouche Riparian Association. "It puts the whole question of riparian rights in jeopardy. If we can't control their fishing, can we control their access to the river? What's the situation with our private property rights? Can they be charged with trespassing? If the rights of riparian owners aren't there anymore, if private property means nothing, how long is it going to be before [the owners] pack it up and say, 'What's the use?'"

The following facts were agreed upon by the Crown and defence at the beginning of the Pelletier trial: The Pelletiers were fishing using gear considered to be illegal (for non-natives) on a scheduled salmon river; they were fishing in an area that was part of a Crown grant; the Pelletiers are Maliseets, direct descendants of Indians who were parties to the 1725 Treaty of Boston. And the Crown added this crucial concession: If the treaty rights of the Pelletiers had not been extinguished by the Crown grants, then they had the right to fish on riparian water with spinning gear. The question of what rights are attached to the Crown grants and what rights are implied in the treaties is at the heart of the dispute. McCormick made it clear that he was limiting his concessions to the Pelletier case only, but the case clearly has larger implications.

The Crown called only two witnesses — a game warden who was involved in the arrest of the Pelletiers, and Dr. Steven Patterson, a professor of history at the University of New Brunswick. Patterson is a specialist in 18th-century Nova Scotia and New Brunswick history. When Patterson took the stand, the trial turned into a fascinating lecture on New Brunswick history.

In the late 1600s and early 1700s, the British and French were at war over who controlled the land we now call Nova Scotia and New Brunswick. The French were defeated in 1710 and surrendered a large part of Acadia, the portion we now call Nova Scotia, under the Treaty of Utrecht. The natives in Nova Scotia and New Brunswick had sided with the French. The Micmac, the largest native group, occupied land all along the east coast to the Bay of Chaleur; the Maliseet occupied the St. John River valley; and the Passamoquoddy Indians lived in Maine.

In the summer of 1725, a delegation of natives and a representative from Nova Scotia travelled to Boston to discuss a peace treaty with the British. A treaty was signed on December 15, 1725, and was ratified at Annapolis in June of 1726 by 16 Indian chiefs. Under the Treaty of Boston, the natives were to have benefits "equal with His Majesty's other subjects." In return, the natives acknowledged that King George was the "rightful possessor of the province." The treaty made the following statements about property rights: The British promised that "the Indians shall not be molested in their persons, hunting, fishing and planting grounds"; and the Indians promised "not to molest any of His Majesty's subjects or their dependents on their settlements already made or lawfully to be made."

"I think that's a clause that's been frequently overlooked," Patterson told the court. "Some people have suggested that these treaties do not deal with land, but in actual fact there is a general statement in this treaty and all subsequent treaties that says that the Indians shall not molest any of His Majesty's subjects in their settlements already or lawfully to be made. I interpret that to mean that the Indians are to accept that there will be European or white settlement in Nova Scotia and that they're not to interfere with them." ("Molest" should be understood to mean "interfere with," he added.) "They will not be molested and therefore must not molest others. It's clear that neither party is to take the law into their own hands, but they're to rely on the laws to settle disputes within the framework of His Majesty's laws." Between 1744 and 1748 Britain and France were at war again. Many natives, despite the treaty, renewed their alliance with France. The Treaty of Boston was renewed in 1760, and again in 1778 during the American War of Independence.

Meanwhile, the British government was setting aside reserves exclusively for natives. However, the government refused to set aside large uncultivated tracts of land for these reserves. Under English common law, Indians had a right solely to the land where they were located, but the reserve land was not large enough to support a fishing or hunting range. Then white squatters began to move in on the reserve land — with the blessing of the government. If the natives were not to be molested on their hunting, fishing and planting grounds, and if their reserves were too small to support these activities, Patterson said, where were they to go?

*

The Crown prosecutor presented an argument based on evidence about an early land grant to Samuel Lee, a merchant interested in exploiting the salmon fishery on the Restigouche River in 1788. The grant gave him the land, "together with all woods, underwoods, timber and timber trees, lakes, ponds, fishings, water and water courses." The Crown maintained that this evidence showed that exclusive fishing rights went with land grants.

The defence called no witnesses. Before closing arguments, Judge Arseneault put his finger on the case's central question: "It's a question of private lands and whether treaty rights extend over those lands." McCormick said the province's position is that Indians are not allowed to hunt and fish on private land. The treaty states that future settlements are not to be molested. "The treaty itself provides the answer," he said.

Hatchette disagreed. He listed some of his multitude of defences: when the Crown signed the treaties, it promised not to molest natives in their hunting, fishing and planting. "The treaty is to protect Her Majesty's subjects, and I don't think Americans [who allegedly owned the land where the Pelletiers were fishing] are Her Majesty's subjects. We can't just accept that he's got an exclusive right to fish and that's to the exclusion of Indians, because there's no evidence to indicate that the two can't possibly co-exist."

In his decision, Judge Arseneault concluded that the Crown had run into trouble right from the start. He said that to show private fishing rights, the Crown had to show that the property was subject to a grant before 1884, and that it bordered non-tidal waters. The judge said the province failed to bring forward any evidence that the Pelletiers were fishing in non-tidal waters, although they were arrested upstream from Tide Head. He acquitted the Pelletiers on that point alone.

Judge Arseneault said in his decision that because of the "important issues" raised by the case (and in the event that he was wrong about the non-tidal waters question), he addressed the case's substantive issues. "One must remember that the treaty clearly and plainly sets out two competing interests: On the one hand, there is clear recognition in the treaty that the Indians have a right to hunt and fish freely. On the other hand, the treaty, by its very terms, places some limitation on that right by making it clear that the British had every intention to colonize, settle and use the conquered land. The Indians had to have been aware of that. Such being the case, I believe that the parties to the treaty must have intended that, with the passage of time, those two competing rights would have to be reconciled and accommodated where possible."

The judge focused on the weaknesses of the Crown case. McCormick didn't call any evidence about the origin of the Crown grant in question. The riparian owner never appeared in court to give evidence about how the private land was being used and whether the Pelletiers were interfering with the use of the property. "I have not been persuaded that the use of the river by the defendants for fishing prevented the owners of the alleged riparian rights from exercising their rights as owners."

The judge continued: "Did the province of New Brunswick have the constitutional authority at that time to grant lands resulting in the extinguishment of treaty rights? I conclude that, after 1867, provincial Crown grants could not be used to extinguish existing treaty rights as that power had been removed from provincial hands at the time of Confederation."

The judge pointed out that the burden of proof was on the party alleging that a treaty had been extinguished. "It is now accepted that treaties are to be given a fair, large and liberal interpretation in favour of the Indians and that any doubtful or ambiguous expression must be resolved in their favour."

❋

Al Carter says the Pelletier case is a major concern for riparian owners. "If the courts rule that the natives have the right to fish and that their rights supersede our 1880 riparian rights, then we might as well just pack our bags and go home. It's got to be a big concern for Premier Frank McKenna and his people. They lease us rivers and the lease gives us the exclusive right to fish. The so-called rights have not been defined. The government has got all kinds of Crown land with rights — lumber rights, mineral rights, mining rights, fishing rights, hunting rights. When somebody comes along and says you don't have those rights and you can't lease them and you can't sell them and give them, it's got to be a concern."

The Crown appealed the decision. In March, 1996, Mr. Justice John Turnbull upheld the acquittal of the Pelletiers. He said Keith McCormick's "failure to prove that the Crown grant bordered non-tidal waters is fatal to the Crown's case." The judge steered clear of making a definitive statement about the rights issue, although he suggested that Crown grants "being later and specific would take precedence" over treaty rights. He concluded: "Any settlement of 'a rights claim' would have to cover all who have aboriginal or treaty rights in New Brunswick if the settlement is to be comprehensive" — an ambiguous outcome in a case that grew to be far more important than David and Rhonda Pelletier's decision to cast their spinners into the Restigouche River.

In the spring of 1996, Rhonda Pelletier was back in court charged for fishing Atlantic salmon in the same Restigouche River pool with the same type of spinning gear. And so the collision of rights continues and important legal questions remain unanswered.

CHAPTER SEVEN

Conservation

Wilfred Carter

I have laid aside business, and gone a-fishing.
— Izaak Walton

Chamcook Stream doesn't look like much. Narrow enough to spit across, its waters run out of the woods, through what was once a farmer's field, under a paved road and down over a bank into a narrow tidal cove in the Bay of Fundy just north of St. Andrews, New Brunswick. Chamcook Stream doesn't look like much, but its waters created Dr. Wilfred Carter's Atlantic salmon miracle. This miracle helps to define the man who has done more to conserve the Atlantic salmon than anyone else in the world.

In 1968, when Carter was the chief salmon biologist for the government of Quebec, he was hired by a group of prominent anglers to be the first president of the International Atlantic Salmon Foundation. The foundation was organized by anglers who owned private fishing waters in New Brunswick and Quebec. Concerned about declines in Atlantic salmon runs, they were willing to put up the money for a first-class international salmon conservation and research organization.

A search committee sought out Carter, who is a passionate conservationist and fine scientist, a gentle, soft-spoken man who can be tough as nails when he needs to be. Carter took the job and set up shop in a cottage in the backyard of his Gaspé home, with his wife Pauline as the foundation's secretary and only other employee. He travelled constantly, promoting the new foundation.

In 1971, Carter decided that his hometown was too remote to be the headquarters of the organization. He looked at Ottawa, Halifax and Fredericton, and then he decided on St. Andrews because it was close to both the United States border and the renowned aquatic research centres in the town. In the area around St. Andrews, Carter searched for a source of fresh water that flowed into the ocean so that he could start a salmon genetics research program. He wanted to raise young salmon in the stream, tag them, release them into the ocean and collect them when they returned to home waters.

Dr. Wilfred Carter, the world's leading salmon conservationist.
ATLANTIC SALMON FEDERATION

When he saw Chamcook Stream, he knew he had what he needed. He bought the farmer's field through which the brook flowed. "We needed a brook or a small stream that salmon could home back to," Carter says. "Every salmon comes back to its own individual river where it was born, and rarely, if ever, does it make a mistake and go to another one." He designed the genetics research centre with $2-million drawn from private sources. They raised young salmon in a hatchery, using the waters of Chamcook Stream. In the spring of 1974, Carter and his staff put their first crop of salmon smolt out to sea.

Local people weren't impressed. No salmon had ever lived in Chamcook Stream, and, Carter says, people thought he was crazy. "I kept hoping that the following year there would be some return of those smolt. People were making some not exactly complimentary remarks about this operation out in the field at Chamcook because they didn't think the salmon would come back to us. I was quite worried about that, because my reputation was on the line. Not only that, but we had spent about $2-million here developing this hatchery for a genetics program that depended on having salmon come back. We planned to cross-breed different strains of salmon, put them in the brook and then send them out to sea. The survivors would come back and then we would be able to measure survival rates and certain strains that were better than others, certain strains that grew faster than others. In order to do that the salmon had to come back."

In the spring of 1975, Carter found himself standing on a point of land halfway along the cove, looking out in the bay, waiting for his salmon to return. "I was pretty nervous the first year, and I kept coming down here almost every day and I'd look out to sea about the time the salmon were supposed to be coming back. It was sort of a fervent prayer more than anything else that some salmon would come back."

On a morning in early July, it happened. "All of a sudden, I couldn't believe my eyes — a salmon, a beautiful silver salmon, jumped right clear out of the water and splashed back down in. I was so thrilled when I saw that. It was a very emotional moment. I think I wasn't far from tears when I realized, yes, it's going to work after all." Thousands of salmon have come back to the brook since that first one, but "that first one was the most important one of all."

Every salmon river should be nurtured in the way Carter cared for Chamcook Stream. Carter says:

> Eventually, if we really want to manage properly and successfully, we're going to have to go to river-by-river management, because every river produces its own stock of salmon and every river has slightly different problems. The management requirements are different, too. You can't successfully manage salmon on a regional or national scale. You have to take a river system and manage that river system like you would if you were a farmer. You'd manage it so you weren't harvesting more of the crop than you should. You'd ensure that you were leaving enough for natural seeding. That's why you've got to get rid of nets in order to do that kind of management because nets are indiscriminate fishers — they harvest everything that comes by. Ideally, you should only be harvesting salmon after they come in to the mouth of the river. And then you should be harvesting them on the basis of how many of them you can safely take without harming the stocks.

Individual river management will mean that governments will have to hand over their authority to local management groups. They can't expect river managers to do all the work without the power to make decisions. Carter says governments must have faith that responsible people can get together and decide what's best for the salmon.

*

Wilfred Carter was born in the town of Gaspé in 1926. His father was a salmon fisherman and the family had a summer cottage on the banks of the Dartmouth River. In those days, the river was leased to a group of Montreal businessmen for private fishing. Young Wilf and his father would sneak down and fish the pools early in the mornings before the wardens arrived and late in the evenings after they left. He got his first taste of salmon fishing there.

His mother's father was a commercial salmon fishermen and he learned another philosophy of salmon fishing through her family. The commercial fishermen felt they had a priority right to the salmon fishery "because it was their bread and butter." They saw the sport fishery as a frivolous amusement for wealthy people. But it was also "bread and butter for the people in the sports fishing industry who depended on it, the guides and the cooks and the wardens and the fly tiers and all the rest of it. It was every bit as much their bread and butter as it was the commercial fisherman's bread and butter, but somehow that link wasn't easily understood or easily explained to commercial fishermen."

As a young man he learned a lot about salmon biology as he fished the famous rivers on the Gaspé — the Dartmouth, the York, the St. Jean. He served overseas during the war, and then returned home to study biology at Mount Allison and McGill universities. After Carter graduated from university he went to work for the Quebec government. He stayed for 18 years, eventually becoming the chief salmon biologist, responsible for all the salmon rivers in the province.

"I am a scientist by profession and a naturalist by inclination and desire," Carter once said. Soon after he graduated from McGill, he received a letter from biologist A.G. Huntsman who gave him this advice: "Observe what animals do, where and when, and soon you will know more about biology than has ever been written." Carter took this advice to heart. He described the benefits of observation in an address to a salmon conference in the 1970s: "Sit and watch salmon attempting to leap a waterfall, and note carefully their route as they jump. Then you can prevent an engineer, who may think he knows more than a fish, from building a costly fish ladder at the wrong location. Watch salmon abandon a long-travelled channel to continue their upstream migration against the opposite bank. They may be telling you that their sensitive noses have detected a pollution source. My grandfather, who knew nothing about biology but who was a fine sailor and very wise in the ways of salmon, told me long before we suspected how salmon navigated that they were guided by the stars. 'Each year that the ice cover off Newfoundland is late breaking up,' he said, 'I have noted that the salmon coming to my nets are delayed. It can only be that they could not observe the stars.'"

*

In 1982, the International Atlantic Salmon Foundation merged with the Atlantic Salmon Association, a Montreal-based conservation group that had been operating since 1948. The Montreal association had been doing what it could (most notably publishing the *Atlantic Salmon Journal*), but it had only a small membership and not a lot of funding. The two groups came together and formed the Atlantic Salmon Federation, which is now the world's leading salmon conservation organization.

Carter led the federation until he retired in 1989; he still maintains an office at federation headquarters and is active in the conservation movement. He says the salmon faces threats today as serious as those on the day he began his work with the foundation. "The disappointing thing to me is that, with all of the curtailment of the commercial fishery and the catch-and-release fishery that anglers have been practicing for quite a while, we're still not seeing a resurgence of salmon stocks as we expected. I find that very disturbing. But I find it also perplexing because we don't really know why. The ocean doesn't seem to be producing as many salmon now as it used to."

Carter wanted his federation to have a large, active grassroots structure so people could participate and understand why it was important to keep salmon in the rivers. "Even if people didn't ever fish them, the fact that salmon were there indicated that that river was pretty healthy." He set up seven regional councils, which made the organization much more than a conservation voice for rich American anglers. He has also promoted a broad educational problem to get his message across: "Keep rivers healthy, healthy enough so that salmon run in them, because that will have a direct impact on our ability to survive as people."

If salmon are to be properly managed in the future, Carter looks for a combination of private ownership and individual river management to do the job. "Private ownership historically has given better stewardship of salmon rivers, and better management. It's limited the pressure on the fishery." The old type of public fishing — the belief that a river belongs to everyone and everyone has the right to fish — "just translates into everybody's fishing being nobody's fishing. You end up with nothing."

Today the Atlantic Salmon Federation is a $4-million-a-year conservation organization with a staff of 50 people. The new president is Bill Taylor, a young conservationist Carter hired in 1988 to work as a co-ordinator with the federation's regional councils and affiliate organizations. The federation raises its annual budget through donations and government contracts. About half the budget comes from private donations, 25 per cent from foundations and corporations, and 25 per cent from government.

The federation's headquarters has two main buildings, a fish hatchery, and a public salmon aquarium and museum beside Chamcook Stream. One of the main buildings is called the Joseph Cullman Building, the other is named after Francis Goelet, two major contributors to the federation. Two powerful men co-chair the federation: Lucien G. Rolland, the former chairman and chief executive officer of Rolland Inc., is the Canadian chairman; and Donal C. O'Brien Jr., the president of the Rockefeller Trust Company and chief legal counsel of Rockefeller Family and Associates, is the United States chairman.

Sixty per cent of the federation's donations come from the United States and 40 per cent from Canada, and the membership of 6,000 is approximately half Canadian and half American. Through its regional councils, the federation represents about 140 conservation and angling organizations that have a combined membership of about 40,000, and which are the force behind the federation's conservation work on the rivers. The federation's voice, the *Atlantic Salmon Journal*, has a circulation of about 10,000.

The federation is involved in both salmon research and public advocacy for conservation on rivers. "Unless we are able to do a good job telling the world that the Atlantic salmon is a barometer of a healthy environment, and unless we are able to gain large-scale support for what we're doing, not just from salmon anglers but also from the conservation community at large, we've got a tough battle ahead of us," Bill Taylor says. "We need more members, we need more money so that we can expand our programs, and we need to do a better job selling ourselves and the work we're doing so we can attract that support. What we're doing affects the environment in a much broader way than many people believe."

Bill Taylor, the president of the Atlantic Salmon Federation. ATLANTIC SALMON FEDERATION

✻

Izaak Walton's *The Compleat Angler*, which is much more a discussion of the nature of human virtue than a book about fishing, will provide Carter's epitaph: "I have laid aside business, and gone a-fishing." Carter still loves to fish, but the sport has changed for him over the years. "I've reached the point in my life where going fishing is more important than catching fish. When you reach that point, then you really understand what fishing is all about. It's not about catching fish. It's about being out in a beautiful place with pleasant companions, enjoying nature, seeing other wildlife and game birds, and participating in the natural system of the outdoors. I now feel good that I can get more pleasure out of going fishing than actually catching fish. I don't catch very many fish any more. I'll fish for a little bit and then I'll sit on the bank and think about some of my problems or just enjoy the beauty of the scene."

Carter was awarded an honorary doctor of science degree from the University of New Brunswick in 1979, and in 1993, he was given an honorary doctorate from Mount Allison. That same year, he received the Order of Canada. "His contribution has been immeasurable," says Dr. Alex Bielak, who worked for Carter at the federation from 1984 to 1990. "A lot of people ask him for advice and a lot of people respect that advice."

"The ups and downs of the salmon world could always be measured with the greatest accuracy by Wilfred's mood," says Jack Fenety, a friend of Carter for more than four decades. "A smile noted that all was well, while a gruff glance showed his displeasure."

Carter's glance today is gruff, and he is less optimistic about the salmon's future than he has ever been. "I have colleagues and friends who say it's just part of the cycle and they will come back. I don't agree with that. I won't believe it's just part of the cycle because there's so many other factors involved now that weren't there in earlier times. I guess I'm less optimistic in some ways, but then mankind is really pretty damned resourceful. When he gets to the point where he sees his own very survival is threatened, I think that the things that have to be done will get done. Not only will man survive, but salmon and other creatures which make mankind's life meaningful will survive too."

Jack Fenety

We are surrounded by, and soon to be taken over by,
our own sins and those of preceding generations.
— Jack T.H. Fenety

In the fall of 1995, in a cedar-panelled room above the Woodsmen's Museum in Boiestown, New Brunswick, the missionary of the Miramichi rose once again to preach to his converted flock. "We need all the help we can get," said Jack Fenety in his annual address to the Miramichi Salmon Association. "We, together, all of us, must seek out the many hundreds, yes even thousands, who fish or at some time fished the Miramichi. I truly believe each of the thousands out there do in fact owe the Miramichi a debt for countless hours of enjoyment. Our struggle is far from over. We, as salmon conservationists, must continue to practise and support our beliefs. We must continue to make the sacrifices still required to fulfill our founding dream of 1953."

After more than four decades working for Atlantic salmon conservation, Fenety was launching his final campaign as president of the association — a plan to raise more than half a million dollars for the Miramichi River system. The dream of 1953 — the year the Miramichi Salmon Association was formed — is to restore the Miramichi's Atlantic salmon runs to numbers that equal the great runs of the past.

Fenety has been spreading the same message since the association was founded: We all must become actively involved in conservation; if we want to save our rivers, we have to do it ourselves. In his early speeches as president of the association, Fenety used to say that we can't afford to lose "even one puddle" to pollution. The fiesty president told the membership: "There can be only one way out. The power to eliminate the licence taken by the polluters, be they large or small, rests with you and me. We the people are the greatest power in North America. We must harness this latent power and put it to work, and time is of the essence. Seek a communion with your conscience. Decide yourself if you are a big, fat, complacent cat. Examine your position. Are you now operating or working for, or do you hold shares in companies whose operations, if continued in their present fashion, will lead to the ultimate destruction of the country you love?"

Fenety, 76, preaches like an evangelist about his favourite subject. In a 1968 address to the Miramichi Salmon Association, he said, "We can be justifiably proud of our achievements in many fields and in many directions — freedom of speech, freedom of worship, freedom of assembly and hundreds more. But I ask you, under what constitution, under what bill of rights, have we been granted the freedom to destroy — and that's what we're doing — destroying our human heritage of clean water, clean air, clean land and the fish and the animals that go with them. Our conduct toward our God-given renewable, natural resources has been, in plain words, disgraceful. Not only that, our rampage of desecration continues at an accelerated pace. Let all of us stand up for what is right — and conservation in all its forms is right."

Today, Fenety still calls for an "'Onward Christian Soldiers' type of campaign" to convince people to contribute more to conservation. He has been president of the Miramichi Salmon Association since 1961, and during that time one of the oldest conservation organizations in North America has been one of the most effective. To this day, his association has lived up to the Fenety ideal, paying its own way, accepting no grants from governments.

⁕

Like a prophet, Fenety advocated an end to the Maritime commercial salmon fishery years before it became a contentious public issue. He demanded the closure of the Newfoundland salmon net fishery, which he believed, rightly, was intercepting fish returning to rivers all over North America, delivering his first speech against that fishery 37 years before Ottawa finally closed it. He was fighting the budworm spray program in the 1950s. Years before the creation of the

North Atlantic Salmon Conservation Organization, he suggested that conservationists create an international body to help manage the Atlantic salmon.

About many of these important issues, Fenety now says: "I said they would rue the day, and they did rue the day." He has always been a force that politicians have had to reckon with. He once called for Romeo LeBlanc's resignation when the then-federal fisheries minister said that the problem with Atlantic salmon stocks lay not with netsmen but with anglers. Fenety hit the roof and wrote to LeBlanc: "You can ignore Jack Fenety, but you ignore the Miramichi Salmon Association at your peril."

Joseph Cullman smiled when the name Jack Fenety was mentioned during a conversation at his Runnymede fishing lodge on the Restigouche River. "When Jack Fenety has spoken, is there anything left to say?" he asked. In February 1996, Cullman gave Fenety $25,000 for his river campaign.

❋

Jack Fenety grew up in Fredericton. His father, a car salesman, was an outdoorsman who took his two sons trout fishing and hunting on the weekends. Fenety grew to love the New Brunswick wilderness, and even within the city limits of Fredericton spent time in the University of New Brunswick woodlot, snaring rabbits and weasels.

In the 1940s and 1950s, anglers, riparian owners and outfitters on rivers all over Atlantic Canada began to notice a decline in the salmon runs. In 1953, a group of concerned anglers formed the Miramichi Salmon Association, which, they decided, should have an international membership because many Miramichi anglers lived in the United States. When the association was incorporated, its rules established that its president, the majority of the directors and the secretary-treasurer would be Canadian. Control of the organization would remain in Canada, but anglers and property owners in the United States would have a strong voice.

When the association was formed, Fenety was a working journalist, a radio broadcaster at CFNB in Fredericton, not a riparian owner or even a salmon fisherman. Malcolm Neill, a boyhood friend of Fenety's and the owner of a Fredericton sporting goods store whose family owned CFNB, was asked to join the new Miramichi Salmon Association. He brought Fenety with him. Fenety first became the organization's public relations director, then a director, then president. "They call me the president for life. I think that nobody else wanted the job."

In his early days at the association, Fenety's target was the budworm spray program and the pesticide DDT that was killing millions of baby salmon each year. When the budworm spray program began in the early 1950s, Fenety was hired to report on its progress for CBC and CFNB in Fredericton. Impressed by the positive results of the first year of the program, Fenety narrated a film that was shown all over Canada, extolling the virtues of DDT. Then he realized that the pesticide was killing birds, fish and many other forms of life in the watershed. He immediately turned and became a leader of the anti-budworm spray movement.

*

Fenety's Miramichi Salmon Association has always been about more than tough rhetoric and advocacy. There were practical matters to be addressed. The association hired and outfitted two wardens, provided them with boats and motors and paid their salaries. These were the first uniformed wardens on the Miramichi River. When the association was formed, there were few limits placed on anglers, who could kill as many salmon as they wanted. The association pushed for limits and got them. The limit has gone from 21 salmon a week down to the current limit of eight grilse a season.

Fenety recalls promoting catch-and-release of large salmon in the days when putting a salmon back in the water alive was unheard of. He remembers catching a fish one day, watched closely by an old man on the bank. When Fenety released the salmon, the man was amazed. "My God, sir," he said, "you let that go. That fish would have been just the thing to chase down a nice New Brunswick potato."

The association proposed that certain areas of the river be closed to create a sanctuary for spawning at a time when there were no closed waters anywhere in New Brunswick. Subsequently, an agreement was reached with governments and private landowners, and 40 miles of headwater streams were closed. In the 1960s, Fenety began lobbying for tagging of salmon; in 1979, when salmon conservationist Bud Bird was New Brunswick's minister of natural resources, tagging laws finally came into effect.

Fenety also lobbied for salmon conservation in the ocean, and he pushed Ottawa to limit the salmon-netting season in New Brunswick. "I suggested at that time, if the government of Canada could pay thousands of farmers in the west to let their ground lie fallow, we could certainly afford to pay a few hundred commercial fishermen in New Brunswick not to fish for salmon."

In the 1960s, Fenety asked the federal government to study the Newfoundland fishery to see if it was intercepting fish from mainland rivers. He didn't get anywhere. In 1975, he finally persuaded then-federal minister of fisheries Jack Davis to investigate. A federal study showed that, in some places, more than 50 per cent of the catch in Newfoundland waters was destined for mainland rivers. At one time there were more than 5,000 legal commercial salmon fishermen in Newfoundland. Then the number was cut to 4,000, then 3,000. "All that meant was you had fewer people catching more fish." In 1992, Ottawa announced a five-year moratorium on the Newfoundland salmon net fishery and started buying back licences. Now all that remains is the Labrador fishery, which scientists are recommending should be closed.

The association opened an office in Boiestown with a full-time manager and a secretary. Beginning in 1981, Fenety acted as the full-time (unpaid) manager of the association and initiated a number of programs. The association hired a salmon geneticist from Norway who discovered that there were at least 11 different genetic strains of Atlantic salmon within the Miramichi River watershed. The association sponsored postgraduate biology scholarships at the University of New Brunswick. Thousands of parr were being rolled in flour and fried every year by fishermen who didn't know the difference between a trout and a baby salmon, so the association began a crucial education program to teach people the difference. Finally, the association lobbied to have salmon rivers "scheduled" for fly fishing only, a regulation that is in force today.

In February, 1996, in a Boston hotel, Fenety told a dinner audience that he would be stepping down as president. Bud Bird was elected as the new president.

Jack Fenety, the president emeritus of the Miramichi Salmon Association, has been preaching river conservation for more than 40 years.
DAVID RISING

*

The association now has an annual budget of about $100,000. Fenety has launched the fund-raising campaign so it can hire a (paid) full-time manager to replace him. "Our aim and objective was always to assist government and put our money where our mouth is," he says. "You have to be seen to be responsible and financially responsible if you're going to advocate a point of view. You've got to be able to back it up. You have to say, 'I'll raise that money and I'll do that.'"

The association has been putting its money where its mouth is every year with its "satellite rearing" operation, which raises young salmon in fibreglass tanks beside the river during the summer and then releases them in the fall. The satellite rearing program is paid for by association members and the fishing camps where the tanks are located. There were 20 satellite tanks on the Miramichi system in 1995, and each tank puts about 5,000 fish in the river.

*

Miramichi outfitter Vincent Swazey notes that for most of his years in the conservation movement, Fenety had no financial interest in the river — he owned no water and belonged to no fishing clubs. He worked hard "just for the love of fishing and salmon." Fenety called for people to give their time and energies to their rivers, and then he set the example himself. He recently joined the Miramichi Fish and Game Club on the Northwest Miramichi, a famous 100-year-old fishing and hunting club. He says he wouldn't have joined if he wasn't planning to resign as president of the association. "I've given the major part of my life, my mature years, to the salmon. And I don't regret one minute of it. I'm still going to be a conservationist until the day I die."

When Fenety rose to preach to the converted in Boiestown in the fall of 1995, he told them that for the fourth year in a row the Miramichi has reached its spawning targets. "It is a source of great pride and joy," he said. "However, we must remain vigilant. Our stewardship must be firm. Conservation must remain our most important goal."

Lee Wulff

Lee Wulff was to fly fishing what Einstein was to physics.

— Charles Kuralt

Lee Wulff transformed fish-killing Atlantic salmon anglers into river conservationists. The importance of that single contribution to the salvation of Atlantic Canada's salmon rivers can't be overstated. If we are to save our salmon rivers today, we must continue to listen to his words. This most famous of all fly fisherman, who was born in Alaska and lived most of his life in Newfoundland and New York state, fought to preserve the salmon until his death in 1991 at the age of 86.

Howell Raines pays tribute to Wulff in *Fly Fishing through the Midlife Crisis*. "On the *New York Times*, we have a bias against superlatives," Raines writes. "There is a handsomely paid editor in charge of keeping superlatives and other grammatical gambles out of the news pages. Our view is that it is hard to know what is really 'the greatest' or 'the most,' since history is long and our vision is finite and one person's medicine is another's poison. But this is not the *Times*, so I can let you in on a piece of history you can take to the bank. In 1938, Lee Wulff, who was probably the greatest all-round fly fisherman who ever lived, published what is certainly the most important single sentence written about angling ethics in the twentieth century. It was: 'Gamefish are too valuable to be caught only once.'"

Renowned angler Lee Wulff was the inventor of catch-and-release fishing and an important Atlantic salmon conservationist.

Courtesy of Joan Wulff

After Wulff's death, Nelson Bryant, the outdoor editor of the *New York Times*, wrote: "Although he was often dogmatic and unbending and had no fear of treading on anyone's toes, most of those whom Mr. Wulff encountered soon understood that his fundamental concern was for the salmon, that he had no other axe to grind. He was the conscience of the Atlantic salmon restoration effort in North America." American businessman and salmon conservationist Joseph Cullman says: "Lee Wulff was a leader, an icon. He was the one who led the charge to save the Atlantic salmon."

Wulff was born in Valdez, Alaska, the son of the frontier town's deputy sheriff and editor of the *Valdez Daily Prospector*. As a small boy, Wulff fished Alaskan streams that were teeming with trout and Pacific salmon. "My mother used to tell me I was fishing with a bent pin on a piece of string as soon as I could crawl fifty yards to the 'crick' behind the house," he later wrote.

When he was a young boy, his family left Alaska and moved to Brooklyn, New York, and then later to California. He graduated from Stanford University with an engineering degree before going to Paris to study art. He returned to New York City and began to work as a commercial artist. On his days off the athletic six-footer was a regular visitor to New York state's rivers, where he studied the art of fly fishing. He began to invent fly patterns and created a new line of dry flies he named the Grey Wulff, the White Wulff and the Royal Wulff, which became famous throughout the world. He supplemented his income by writing and lecturing about fly fishing.

In the summer of 1933, he travelled to Nova Scotia and hooked his first salmon on Cape Breton's Margaree River. Soon after that he decided to try to make a living fly fishing — giving lectures, writing and fly tying. In 1938, he published *Lee Wulff's Handbook of Freshwater Fishing*. In the introduction he wrote about catch-and-release fishing and included that most important sentence:

Lee Wulff doing what he did best — playing an Atlantic salmon on light fishing tackle.
ATLANTIC SALMON FEDERATION

"Gamefish are too valuable to be caught only once." To say this thought was ahead of its time is an understatement. This was 1938, decades before anglers were thinking about anything but making sure they killed their limit.

The same year his book was published, Wulff was hired by the Newfoundland government to promote sportfishing and hunting in the province. For the next decade, he fished and hunted, made films, wrote stories and gave lectures about outdoor sports. The Newfoundland and Labrador rivers were filled with fish, but as soon as he arrived, Wulff began talking about conservation. The St. John's *Evening Telegram* reported in October, 1938, that Wulff spoke to the Rotary Club about the "absolute need" for conservation of wildlife. "These natural resources are your country's treasures. They must be handled as carefully as any other business to give fair return in sport and in dollars. Your salmon — and I use the word advisedly because they are part of the country and belong to no man or group of men but to every Newfoundlander — are becoming fewer and fewer every year."

During the Second World War, Wulff worked as a recreation adviser to the United States army in Newfoundland, arranging fishing trips for soldiers on leave. In 1944, he opened several salmon camps on Newfoundland rivers and started bringing fly fishermen to the island. To feed his guests at the camps, he put some of the salmon he hooked into ponds where they lived for a couple of days until they were needed for food. This system allowed the camps to survive without ice and also helped to prove that hooked and released fish did indeed survive.

When Wulff was 42, he earned his pilot's licence and flew his tiny single-engine float plane into remote rivers in Newfoundland and Labrador. His wife, Joan Wulff, says: "He had the first small plane in Newfoundland and Labrador so he could fly to rivers no one had ever reached before. He cast the first fly on those rivers, then he wrote about it and did films and showed those films in the United States. And he brought salmon fishermen to Newfoundland. There's no doubt about it, he opened up that country. That's where his reputation was really made because he became so good at everything that he was doing."

Joan Wulff says her husband was obsessed with catching salmon on short, light rods so the game between man and the fish would be as equal as possible. Using such a rod meant he had to do everything just right or he would lose the salmon. "It was all based on making it challenging for him, so that it was skill, not brute force, which is what happens to anyone who gets very good at what they do. He was very good. He knew how to play fish. He had been a filmmaker and in trying to make the fish 'act' before the camera he had to make those fish do something bring the fish to the camera, so he learned how to play them very, very well."

His great angling skill made all the difference. When he spoke about conservation, salmon anglers, who admired him, took his message seriously. In 1958, he published his classic book *The Atlantic Salmon*. The book focuses on salmon angling, but its overriding theme is conservation; this book was written by a man who was making his living as an angler, and he was telling people to stop the killing. The following passage from *The Atlantic Salmon* could easily have been written today. We should take these words to heart.

> One of the greatest needs of the salmon is a sanctuary area near their spawning grounds where they can be safe. When a great fish has made his sea voyages and returned up the long and tortuous way to his spawning grounds, when he has been through the nets, evaded the sharks and otters and eels and seals, and the flies of anglers, it is only reasonable that man, after helping to oppress him so heavily, should give him a place to rest and feel secure.

*

In the summer of 1966, while filming an ABC television sportsman's show on tuna fishing — another type of fishing he had mastered — Wulff met Joan Salvato Cummings, who was also an expert angler and champion fly caster. They fell in love on the trip and were later married. Joan and Lee Wulff both loved New Brunswick's Upsalquitch River, where they fished for salmon every year at Joseph Cullman's Two Brooks camp. Wulff, who fished all over the world, said the highlight of his fishing season was always fishing on the Upsalquitch.

Joan Wulff started fishing for Atlantic salmon after she met Lee, and she soon joined his conservation campaign. She is still working hard to spread the word about salmon conservation. "I think the catch-and-release concept was his most important contribution. The way my husband used to put it was that a farmer would not kill off his best breeders for hamburger. You always have to maintain your stock. A farmer can understand that, and yet they lose track of it when it comes to fish."

In the 1960s, a section of trout water in Yellowstone Park became the first mandatory catch-and-release fishing area in the United States. Twenty years later, catch-and-release came to New Brunswick, when anglers were told they had to release all large spawning salmon. "It really has not developed all over the country as much as it should have," Joan Wulff says. "But the fly-fishing community has embraced the idea and whether the regulations say so or not, it's catch-and-release, which is wonderful. Lee made people aware. He stood up against public opinion, against the opinion of the people he was working with, and talked about salmon from their point of view. When he spoke, people listened. And he was a very good speaker. He was an original thinker. He did not mouth things that other people said. It had to make sense to him and he had a very deep understanding of the natural world."

Wulff became a major player in the conservation movement through his involvement in the Atlantic Salmon Federation, serving as a director and vice-president, chairman and honorary chairman. In its early days, the federation was supported mainly by people who owned private water. They were working hard and putting money into conservation, but Wulff believed the ordinary angler should become part of the campaign, because the strength of the movement would be measured in numbers. In the early 1980s, the federation sent Lee and Joan Wulff on an Atlantic Canadian tour. They held meetings and workshops and talked about conservation, stirring up support for conservation at the grassroots level.

On one major conservation issue, Wulff became deeply divided from his friend Dr. Wilfred Carter. Wulff urged the Atlantic Salmon Federation to push governments to declare the salmon a game fish, a status that would allow it to be fished only by river anglers. While Carter agreed that commercial netting of salmon in the ocean was a great threat to the resource, he thought game fish status was unrealistic and there was more to lose than to gain by lobbying hard for this reform.

The federation, rightly or wrongly, was thinking about politics; Wulff was thinking only about what he believed was right. In *Lee Wulff*, a biography of the famous angler, author Jack Samson describes Wulff's battles with the federation. "Lee had never been enthusiastic about the scientific approach to salmon management. His theory was that the salmon was right there in the river and why try to change it by fooling around with gene research and the like. The St. Andrews project was almost completely concerned with the commercial raising of salmon for food and Lee considered it a waste of time. As a result his tirades against net size and for the game fish status much of the time fell on deaf ears. This did not amuse him."

*

On a routine flight to requalify for his pilot's licence in April, 1991, Wulff's plane crashed and he died. "He had the best of it in terms of salmon fishing," Joan Wulff says. "He saw the whole thing but was working with all his passion at the end to try to get salmon to be declared a game fish. He died thinking about that."

The year before Wulff's death, Carter wrote a letter of reconciliation to his old friend. The letter, which is included in Samson's book, is as good a eulogy as any for Wulff.

> Dear Lee,
>
> I am sitting in my den overlooking the Bay of Fundy on this beautiful spring evening, wondering how far away the returning schools of salmon are, and thinking how lucky I am to be alive and well, and dreaming of the fishing that is to come this season. I am thinking too of you — and selfishly of me — and wishing that as we are growing older that we were growing closer — instead of farther apart, because I have sensed, as I am sure you have too, the disagreements we have had on some important salmon conservation issues have influenced our personal relationship, and I deeply regret that this has happened.
>
> No one has made a greater contribution to fisheries conservation, and to the salmon's cause in particular, than you have, and generations of fishermen will always be indebted to you for your leadership, dedication and yes, for your stubborn adherence to ideals in which you strongly believe. You made many of your greatest contributions when it took the most courage to be innovative, and different.
>
> I really wish we could be good friends while still respecting each other's right to hold differing opinions. In our separate ways we can each claim to have made significant contributions to conservation — yours far outweighing my own. Perhaps, if I am lucky, I will live to be 85 too, but I will never live long enough to even approach the record of your enormous contribution to the welfare of our beloved salmon.

Famous angler Joan Wulff, the widow of Lee Wulff, with Joseph Cullman III outside Runnymede Lodge on the Restigouche River. PHILIP LEE

Runnymede

Joseph Cullman III is one of the most distinguished businessmen in the United States. He is chairman emeritus of the billion-dollar conglomerate Philip Morris Inc., and was chief executive officer of the company for 21 years. A New York native and former navy man, Cullman was twice voted the leading industrialist in the United States.

For more than 40 years, Cullman, 83, has been coming to New Brunswick to fish for Atlantic salmon. He runs two exclusive salmon lodges — one on the Restigouche and one on the Upsalquitch. He has been a major force in the salmon conservation movement and may be the largest financial contributor to the cause. Cullman was one of the founders of the Atlantic Salmon Federation and was chairman of the organization in the United States for more than 20 years.

Cullman says there are three major threats to the survival of the Atlantic salmon today: Nature, which is creating unfavourable conditions for the salmon at sea; pollution in rivers and the ocean; and native net fishing on salmon rivers. "The Indians stand out as the only people who are flagrantly ignoring the principles of conservation," he says. Cullman's deep concern about the effects of native gill-netting of salmon by members of the Mi'gmaq Listuguj Nation at the mouth of the Restigouche River is echoed in lodges throughout the watershed.

"We've been trying to persuade them [the natives] to use other methods of collecting the fish, a trap net rather than a gill net, trying to get them to reduce their take, but so far the Indian fishery has remained adamant that they have aboriginal rights that are theirs and theirs alone. They claim to have fishery rights over the whole country of Canada. The Indians are trying to sell fish that could be bought — aquaculture fish raised differently — without interfering with the native run of fish."

I interviewed Cullman at his Runnymede Lodge in July, 1995. The lodge sits beside the main road that winds along the bank of the Restigouche, through Tide Head towards Campbellton. Runnymede's wide lawn and gardens stretch down to the river. The lodge is named after the 1215 British battle at Runnymede where the barons forced King John to sign Magna Carta, the charter of British liberty, which, among other things, guaranteed hunting and fishing rights.

The Runnymede river property was purchased in 1896. Cullman first came to Runnymede in 1947, and he and his younger brother Edgar bought it six years

Runnymede Lodge on the Restigouche River is the river home of Joseph Cullman III, a leading salmon conservationist. PHILIP LEE

later. Cullman and his brother, who is president of the Restigouche Riparian Association, own 9.5 kilometres of the Restigouche River and lease one small stretch of water at the Rafting Grounds. They began leasing 9.5 kilometres of salmon water on the Upsalquitch River in 1962, and they operate a lodge beside those waters at Two Brooks. Between his two lodges, Cullman can handle about 14 guests.

Cullman is a small, gracious man. He invited me into his lodge's large screened porch, which offers a wide view of the Restigouche River. A strong river breeze filled the room. On the porch were two of his guests — outdoor writer Nelson Bryant and Joan Wulff, who was relaxing on a couch reading when I arrived. Bryant, a large man with a white beard, was sitting at a table working on his fishing gear and smoking a pipe. I turned on my tape recorder and they started talking about salmon. Our conversation took place during the height of the native salmon fishing protest on the Northwest Miramichi River in the summer of 1995. Newspaper articles detailing the tense dispute over the right of natives to gill-net salmon at Big Hole Tract were strewn around the room. The latest stories had described a drive-by shooting on the Eel Ground reserve.

Cullman — who calls himself a long-time Atlantic salmon enthusiast — began the conversation: "The concern began in 1968 when we founded the International Atlantic Salmon Foundation and determined something should be done about saving the Atlantic salmon. On a long-term basis I think the Atlantic salmon has a very good chance of remaining a viable sport fish, although it should be deemed a game fish. We've done so many positive things that theoretically should have increased the numbers of Atlantic salmon returning to the rivers, yet the numbers are relatively stable. But if you look back — and we have the kill record here over the past 25 years — there have been some severe ups and downs.

I think more than anything it has been nature, and the Indian netting, both of which you can't do much about."

Bryant interjected: "The larger philosophical question interests me. There's no question that the Indians in this country and the United States got screwed [in treaty agreements]. But we've reached the stage where the important thing to me is the conservation of the resource. If you conduct yourself in such a fashion as to damage or destroy the conservation efforts, then in the end you have nothing, and the Indian has always been cited as a working conservationist. You know, they didn't clean everything out, they moved around. But it seems to me a point comes when the resource is threatened, when you've got to join hands or you have nothing. And I think that's where we are now.

"I live in the United States on Martha's Vineyard near Cape Cod, and there's a move afoot now to make the striped bass a gamefish. And at first I was against it because I've been a commercial fisherman in my life and I have friends who are, and I said this just shouldn't be a plaything for people who want to catch fish. But I've changed my mind in the last month, because as far as I can see, any regulations that would come up to protect any ocean fish and to maintain commercial fisheries for them have failed. I begin to think that the only way we can protect a species thoroughly, especially one that migrates, is to regard them as a fish for angling alone."

*

The salmon fishing was slow at Runnymede in 1995 because of hot, dry weather and low water levels. Seven people, including Cullman and Wulff, both experienced anglers, had been out on the river for a total of 49 three-hour fishing sessions. They caught two fish. Cullman said with a smile: "You wouldn't say we are devastating the resource."

Wulff said: "The big worry now is that, despite all we have done, the fish are not coming back, and there seems to be something wrong at sea, and it may be that the food chain has been destroyed. I think whatever has been happening has been happening over a very long period of time. It's not just the last few years. This is when we've seen the dramatic drop. But whatever the circumstances were that made that happen, there were probably 20 years when we weren't aware of it. Out at sea, when all these individual river stocks are mixed together, whatever happens to them is indiscriminate. We can wipe out a whole river."

Joan Wulff thinks it may be time to stop killing all salmon (including grilse) for a few years to see what happens to the river runs. Traditionally, the killing of grilse has not been seen as harmful to the resource because these smaller fish are not carrying as many eggs as the large salmon. And most scientists believe that a grilse begets a grilse, that when small males fertilize the eggs of a large female, their offspring will be grilse. Many river people say we should kill every "nasty old grilse" we can get our hands on to increase the number of large salmon. She thinks it is perhaps time to reconsider this management theory.

The night before our conversation, Wulff had caught a grilse and it surprised her. She had resigned herself to not catching a fish. "But I'm still going out and I'm still casting every cast as if I'm going to catch a fish. I'm still loving my surroundings, and that's what it's all about. The fishing was great. The catching wasn't so good. I love the Upsalquitch. That's my favourite place to be to fish for salmon. I love being where fish live. That's where I want to be."

Cullman agreed: "It's an important part of my life. This is something that I enjoy. I enjoy the concept of doing something special for a very special species, the most interesting game fish in the world. And we have this lovely camp. The life appeals to me and the challenge of doing something to conserve the Atlantic salmon is well worth the time and energy that goes into it.

"I remember the first time I fished — I took 14 salmon that week. I fished with a marvellous guide named Elmer Craswell. It seemed rather dull to me, because I was an inexperienced fisherman. I'd throw the line out 30 or 40 feet and he'd say, 'Mr. Cullman, I'll finish for you,' and he'd hook the fish and I'd land them. I didn't think it was very challenging. I didn't know how to fish then." Cullman knows how to fish now. He spends about three weeks a year on salmon rivers, fishing at his New Brunswick lodges and visiting Quebec's Moisie, Grande Cascapédia and Matapédia rivers. "This is my favourite river, this and the Upsalquitch."

Cullman has four large round fibreglass tanks near a brook at Runnymede where in the summer of 1995 he raised 32,000 young salmon fry. The young salmon grew in the cold water running from the brook, and in the fall they were marked and released into the Restigouche in the hope that they'll come back four or five years later as adult salmon, imprinted with the Restigouche River water. How much money has Cullman put into salmon conservation over the years? Modestly, he said he couldn't give me a figure off the top of his head. "I'm a major contributor. I've given a considerable amount of money and I've left money in my will. When you get to be 83, you think about dying."

When I left Runnymede on that breezy summer afternoon, Cullman was thinking more about fishing than dying as he stood on the lawn beside his friend Joan Wulff and prepared for another evening on the water.

CHAPTER EIGHT

Scotland

The defining river

If paradise be a renewal of what was happy and innocent in our earthly days, mine will be some such golden afternoon within the sight and sound of Tweed.
— John Buchan

Growing up in the valley of the River Tweed, John Buchan drew the woods, hills and water into his soul. His father was a man of "wide culture" who ruled the household with Calvinistic discipline. To Buchan's father, "in the words of the Psalms, all things were full of the goodness of the Lord."

Young John Buchan created a biblical world in his mind and set it in the Tweed valley, in the pastoral Border country of Scotland where the history of each town runs red with the blood of men who died in the conflicts with England. "The imagery of the Psalms haunted every sylvan corner," Buchan writes in *Pilgrim's Way*. "More, as I struggled, a confused little mortal, with the first tangles of Calvinistic theology, I came to identify abstractions with special localities. The Soul, a shining cylindrical thing, was linked with a particular patch of bent and heather, and in that theatre its struggles took place, while Sin, a horrid substance like black

The Leaderfoot Viaduct crosses the River Tweed downstream from Melrose, Scotland. PHILIP LEE

salt, was intimately connected with a certain thicket of brambles and spotted toadstools." The legends of King Arthur and Merlin were in every valley. A hollow in a hill was Buchan's entrance to Arthur's sleeping place.

As a child, Buchan lived with his maternal grandparents on a Tweed valley sheep farm. He fished for trout in small burns, catching as many as four dozen fish a day, filling his pockets and threading the fish through the gills with rushes. "The farm stood at the mouth of a shallow glen bounded by high green hills," he writes. "The stream in the glen joined a 'water' flowing from the west, which three miles off entered the Tweed. Those Border shepherds, the men of the long stride and the clear eye, were a great race — I have never known a greater. The narrower kinds of fanaticism, which have run riot elsewhere in Scotland, rarely affected the Borders. Their people were 'grave livers,' in Wordsworth's phrase, God-fearing, decent in all the relations of life, and supreme masters of their craft. They were a fighting stock because of their ancestry, and of noble independence."

In the summer of 1995, I drove south from Edinburgh toward "Tweed," as this river is known in the Borders, with Buchan's words ringing in my ears. On the road, I was overwhelmed by unfamiliar landscapes and towns. When I crossed Tweed waters below the famous Leaderfoot Viaduct downstream from Melrose, I parked the car in a layby at the side of the road and walked back to the river.

When I looked down into Tweed waters, I bridged the gap between New Brunswick and Scotland. I felt that same deep stirring of the soul I always feel when I look into the waters of the Miramichi. It was then that I understood more than ever that this story I was writing about the Atlantic salmon was about much more than rivers and fish — it was about people and history and about the way we choose to live. In the hills above the Tweed lie the spirits of John Buchan's people. "I have never had better friends," he writes, "and I have striven to acquire some tincture of their philosophy of life, a creed at once mirthful and grave, stalwart and merciful."

In 1935, King George V sent John Buchan to Canada as Governor-General. Buchan by then was known as Lord Tweedsmuir, an essayist, poet, historian and novelist, author of the thrillers *The Thirty-Nine Steps* and *Greenmantle*. Buchan brought his love of rivers to Canada where he fished for Atlantic salmon in the Restigouche, Grande Cascapédia and tributaries of the St. Lawrence.

Rivers defined this man's life. Essays about rivers were discovered among his papers after his death; tentatively titled *Valleys of Springs of Rivers,* the series opened with a quotation from Saint Francis: "Praised art Thou, my Lord, of Sister Water, for manifold is her use, and humble is she, and precious and pure." In one of these essays, he wrote: "From my earliest youth I have been what the Greeks call a 'nympholet,' one who was under the spell of running water. It was in terms of them that I read the countryside. My topography was a scheme of glens and valleys and watersheds. I would walk miles to see the debouchment of some burn with whose head-waters I was familiar, or track to its source some affluent whose lowland career I had followed. There was no tributary of the upper Tweed that I could not name, and, when I came to possess a bicycle, the same was true of the adjacent waters of Clyde and Annan, Ettrick and Yarrow. At any time in my childhood I could have drawn a map of my neighbourhood which, so far as the streams went, would have been scrupulously exact. Ever since then I have translated every land with which I was connected into the speech of rivers."

*

The whole of the Tweed valley is organized around Atlantic salmon waters. The river is 96 miles long, rising from springs near the source of the famous salmon waters of the Annan and the Clyde. Like New Brunswick's Miramichi, the Tweed has wide-ranging tributaries — among them the Ettrick, Yarrow, Till and Whiteadder, all fine Atlantic salmon waters — draining 2,000 square miles of Border country.

Salmon enter the Tweed every month of the year and the river offers the finest fall salmon fishing in Scotland. Fish spawn up and down the river, penetrating every tiny burn. About 14,500 anglers fish the Tweed every year, and there have been salmon fishermen on the Tweed for centuries. In the Roman ruins at Trimontium, near the town of Melrose, a bronze fish hook has been discovered that dates from 600 B.C.

The famous Dryburgh and Melrose abbeys, whose spectacular ruins attract thousands of tourists each year, were built within the sight of Tweed salmon water, and the monks no doubt left their cloisters to seek fish for their kitchens. The remains of "yairs," the ancient fishing traps of the monks, can still be seen along the river system. At the town of Kelso, the Teviot runs into the Tweed at the Junction Pool, the most prolific pool on the river. From Kelso, the river runs into the sea at Berwick-upon-Tweed, which is across the border in England.

During the Industrial Revolution, the Tweed was hurt badly by pollution and dams. Mills that produced the famous cloth that bears the river's name lined the banks. The Tweed also lost many miles of salmon water to huge reservoirs that supply water to Edinburgh. But a series of progressive salmon acts, pushed forward by the owners of salmon water, improved the conditions of the river and saved the Tweed salmon. Today, the river is watched over by the Tweed Foundation, one of the most forward-looking river management groups in the world.

Now, the main threats to Tweed salmon are the poor environmental conditions in the ocean and the menace of drift nets off England, Ireland and Wales, which fishery owners along the Tweed are campaigning to stop. A campaign flyer reads: "God Alone Can't Save the King. The wild Atlantic salmon, the 'king of fish,' needs your help. Known as 'walls of death,' deadly British drift-net fisheries have become an annual slaughter, a public embarrassment and a national disgrace. They are devastating our wild Atlantic salmon stocks, killing more than a hundred thousand salmon and tens of thousands of sea trout each year."

⁕

Sir Walter Scott was born in 1771 in Edinburgh but spent most of his adult life in the Borders. The great poet and novelist, the author of *Waverly*, *Guy Mannering* and *Rob Roy*, traced his romantic vision to the Tweed valley. His novels describe Tweed salmon waters, and he writes about night fishing with spears and torches. A hundred years later, as a young man, John Buchan fell into the company of salmon poachers like these and recalled the experience vividly.

> The thing was in defiance of the law, and the water-bailiffs were always on their trail, so that there were many affrays by the waterside which ended in broken heads. I have assisted in "burning the water," which was a favourite pursuit of Sir Walter Scott's, and apparently in his day not illegal.
>
> It was an exciting business, for the shallows would be reddened by torches made of barrel staves dipped in tar, and wild figures with a three-pronged fork, called a leister, speared the fish as they blundered upstream. Once I was arrested by the bailiffs with the others, but in consideration of my youth was released with a hearty kick! The story

Scott's View, above the valley of the River Tweed.
PHILIP LEE

got about and was remembered many years after when I stood as a parliamentary candidate for the county. I was assured that I had the poaching vote to a man!

Scott built his Abbotsford mansion on the banks of the Tweed, three miles west of Melrose. There he made a garden that runs down to the river. At Bermerside, on the road to Abbotsford, Scott often stopped to look down on the Tweed, flowing deep and gentle through a thickly wooded valley. His lookout place is now known as Scott's View.

In 1832, Scott died while listening to the sound of the murmuring Tweed. His biographer, John Gibson Lockhart, who was married to Scott's eldest daughter, described his final day at Abbotsford. "About half-past one p.m. on the 21st of September, Sir Walter breathed his last, in the presence of all his children. It was a beautiful day — so warm that every window was wide open — and so perfectly still, that the sound of all others most delicious to his ear, the gentle ripple of the Tweed over its pebbles, was distinctly audible as we knelt around the bed, and his eldest son kissed and closed his eyes."

As Scott's horses drew his coffin towards his grave at Dryburgh Abbey, on a "day dark and lowering, and the wind high," they stopped out of habit at his lookout at Bermerside. The funeral procession stopped with them and gazed across the Tweed to the Eildon Hills.

*

Shortly after 8 a.m. on a bright sunny day in the Borders, Duncan Glen was giving me a lesson in Scottish driving as we hurtled down a narrow, twisting road in a battered old Suzuki four-wheel-drive toward the headwaters of the Tweed: accelerate as fast as you can on the straight stretches, brake hard when you approach one of the 90-degree curves, steer with the reflexes of a cat and hope for the best.

The land in the Tweed valley, one of the most fertile areas of the United Kingdom, is farmed intensively. But the landscape flying past my passenger-seat window was changing all the time — one minute flat fields, the next, steep hills suitable only for grazing sheep. We were headed for the Holmes Water in the headwaters of the Tweed system, where scientists from the Tweed Foundation are doing field research on salmon and trout populations.

Duncan Glen, a native of the Tweed valley who likes to fish for sea trout at night on the River Till, is a habitat improvement manager for the foundation and my tour guide for the day. The Tweed Foundation is one of the best examples anywhere in the world of what a first-class individual river management program can do. Every stretch of water, no matter how small, is important. Every leaf that falls

Employees of the Tweed Foundation conduct electrofishing experiments in the headwaters of the River Tweed. PHILIP LEE

into the current of a tiny burn is valued because it becomes part of the river's ecosystem. Every small improvement to the watershed is a step forward.

The Tweed is a special case among Scottish rivers. Private Salmon District Fishery Boards manage all Scottish rivers other than the Tweed, which has its own acts of Parliament — the Tweed Fisheries Acts of 1857, 1859 and 1969. These acts give the River Tweed commissioners the responsibility for the protection and enhancement of salmon and trout in the Tweed, an area of jurisdiction that covers 2,000 miles of fresh water in the watershed and extends five miles out into the sea. The board of commissioners is made up of 43 appointed members (23 from angling clubs and 20 from independent interests) and 38 members elected by riparian owners. The commissioners charge all fishery owners an annual fee per salmon and sea trout taken from their waters. The Tweed Foundation was created by the commissioners in 1983 to carry out the scientific and habitat improvement work needed to preserve the river. The foundation is developing a sustainable fishery management plan for the entire watershed.

By mid-morning, Glen and I arrived at the Holmes Water, a tiny stream in the middle of a remote sheep pasture, where Dr. Ronald Campbell, the foundation's chief biologist, was conducting an electrofishing experiment. Electrofishing stuns fish so they can be netted, removed from the stream, measured and put back in unharmed. Tony Coleman, the superintendent of the Tweed bailiffs, was leading a crew in the stream, and Ronald Campbell was measuring fish in the back of the foundation truck.

Since Campbell started working in Tweed waters in 1990, he has been collecting data on the number of small fish the watershed is capable of supporting. In a short section of the Holmes Water, he counted more than 100 small fish, several of which were tiny salmon. Over the years he will return to this spot to see how the fish population has changed. In his first five years with the foundation, Ronald Campbell has been establishing the Tweed's data collection systems. "We have to get grips on the key data that we need. How many fish have been caught? How many females do we need to spawn successfully? What kind of stock do we have coming back?"

*

The Tweed commissioners are funded entirely by anglers and salmon netsmen. Everyone has to pay the river owners for the right to fish. Fishing for trout costs as little as $25 a year; fishing for salmon ranges from $25 a day to as much as $110,000 a week for six rods at the famous Junction Pool at Kelso. For every salmon caught on the Tweed, riparian owners pay $70 back into river management; they pay about half that amount for a sea trout. "No government money goes into the management of the salmon stocks," says Campbell. The capital value of the Tweed salmon fisheries is between $125 million and $250 million. Salmon angling brings big money, more than $10-million a year, into the local economy — a Tweed salmon caught on a rod is worth more than $1,000 to the community.

The main business of the Tweed commissioners is protecting fish stocks. About 10 per cent of the commissioners' budget goes towards the Tweed Foundation, which is expected to find ways to raise money on its own. The foundation now has about $500,000 in the bank and operates with an annual budget of about $325,000. The staff at foundation headquarters at Drygrange, near Melrose, includes a biologist, an assistant biologist, two habitat managers, and a director.

One of the habitat managers, Andrew Veitch, works in the streams removing obstacles for spawning fish. The other habitat manager, Duncan Glen, works on the riverbanks, improving stream conditions. "Part of our work is to raise the profile and the understanding and the enthusiasm for what we're doing," says Judith Nicol, the director of the foundation. "And we're really only starting to do that seriously. It's beginning to snowball and it's quite exciting the way it's working."

*

Duncan Glen hates sheep — with good reason. Overgrazing of sheep in the Tweed valley is the most serious threat to fish habitat in the river system. "People look at that and say, what a wonderfully beautiful, natural landscape, when it isn't in the slightest. This has had centuries of grazing on it." Sheep graze right down to the riverbank and destroy all the trees and tall grasses that provide shade and cover for fish. Their grazing also allows the banks to erode, causing the streams to widen out and become too shallow.

Glen is responsible for correcting these problems. The foundation is fencing the banks of the streams and burns to keep the sheep back. It is also planting new trees and bushes along the banks to mix shade and open water. Because the land has been overgrazed for years, there is no seed base for plants to grow back, even after fencing. Glen is reintroducing more than 10 varieties of aquatic plants, native to the Tweed watershed, into the rivers. These plants will trap silt, help stabilize the banks, and narrow the channel, creating better spawning grounds. "This is all soft engineering rather than hard engineering, which involves rock and iron and that kind of thing."

During our tour we stopped at Douglas Burn, a wide, shallow stream with badly overgrazed banks. "This is a real challenge, this one," Glen says. "That water will be like a bath in there today. No cover. No deep pools." Here, the foundation is planning some hard engineering, using current deflectors and barriers in the stream to try to stop the shifting of gravel and constant erosion.

There are few areas of native forest in Scotland. In the Tweed valley, monoculture, single-age forests line the few treed hillsides. The planted forests tend to choke out the small streams, growing right over them and creating a tunnel-like effect. The foundation is encouraging more suitable planting of forests, leaving buffers around small burns with leafy trees like willows to hang over the water.

To complement the work at streamside, the foundation is also removing obstacles from the river. Andrew Veitch is responsible for dealing with the 1,500 or so obstacles that keep fish from migrating upriver in low-water conditions. These obstacles include the remains of old fishing weirs, abandoned machinery, bad culverts and poorly designed bridges. As each obstacle is removed, more river spawning area opens up.

"At the end of the day it's not just about protecting the rights of the fisheries owners, it's about protecting the salmon themselves," Glen says. "If you get people taking out a whole breeding stock from a burn, it takes an awful long time for that to reinstate itself."

Tony Coleman leads the river's campaign against poaching, which continues to be a major problem in Scotland, especially on remote rivers. He and the 11 other full-time bailiffs are hired directly by the commissioners. Poachers net the lower part of the river. And in low-water conditions in the fall, they use gaffs and large, illegal hooks to kill fish. "Of all the poaching forms, that's got to be the worst. Because for every one that you get you're maiming two or three and generally you're taking them off the spawning beds."

He is also fighting against the poisoning of rivers, which is very difficult to detect. In the summer of 1995, poachers put cyanide in the waters of the Carron, a prime salmon stream, and killed 1,000 fish. Scientists believe it will take the river 30 years to recover and the damage to the river will cost the local economy millions of dollars. The penalties for poisoning a salmon pool are harsh, and include imprisonment in serious cases. Most poachers receive fines ranging from $150 to $1,500. "It's a constant battle. There's no way it will ever end," he says.

*

One of the challenges for Duncan Glen is to find out who owns the tiny burns in the headwaters of the system. The Holmes Water, for example, is owned by many different farmers. Large estates in the valley have a number of tenants. Glen works through two tiers of management to reach an agreement on habitat improvement. He finds that landowners are becoming more co-operative all the time. "They're starting to realize that they can't go on just harvesting and taking. They've got to look at putting something back."

The big landowners on the Tweed are the Dukes of Buccleuch, Roxburghe, and Sutherland, and the Wemys and March estates. The foundation has received a great deal of moral and financial support from them. The aristocracy is more inclined to improve river habitat because its primary concern is to leave something behind for future generations. On the other hand, the businessman farmer sees fencing off the river to keep the sheep out in terms of lost pasture, and the small farmer is thinking about getting his stock to the water. "The last thing I want to do is start imposing our ideals of what we want to do on the agricultural community," Glen says. "If we do that, then this project won't last long-term."

*

Ronald Campbell has been marking fish in the estuary with radio tags, trying to determine which runs of fish go to which part of the Tweed watershed. Tagged spring fish have a choice: They can turn right into Upper Tweed, or left into Ettrick and Yarrow. "We've never yet had a tagged spring fish that has gone into Upper Tweed," Glen says. The foundation is trying to find out if there is one Tweed stock of fish, or if these different runs have distinct genetics that determine their age when they migrate to sea and return, and where and when they spawn.

The genetic diversity in the system makes restocking programs difficult. There is no hatchery on the Tweed — it is the only major river in Scotland without its own hatchery. Ronald Campbell says poorly planned stocking programs can harm the genetic identities of the fish. And too many young salmon can lead to a shortage of large salmon. "You've got to know what is the carrying capacity and you've got to be quite certain, absolutely rock-hard certain, that the number of fish is below the carrying capacity. In other words, you need spare food and space there before you stock, because if you stock it on top of the full carrying capacity, you will end up with fewer fish migrating out to sea than if you hadn't stocked at all."

Campbell says salmon river managers can't sit back and blame poor fish runs on problems in the ocean. "In our view, the more salmon we put out the more we can produce. It's no good for people who run the fresh-water part of the cycle to simply say it's all the fault of what's happening at sea, there's nothing we can do. The fishermen and others who work in the sea will just say, 'Look at all the problems in the fresh-water environment.' And the two sides just blame each other and the whole thing spirals downwards."

The catch statistics on the Tweed, which date back to the early 1800s, show major fluctuations. For example, there were enormous catches in the 1860s and 1960s. "It's not hugely different from what it was 50, 60, 70 years ago," Ronald Campbell says. "There've been much bigger peaks. There have been periods of enormous productivity. But there's also been periods of somewhat similar levels to this in the past." For 25 years, about 10,000 Tweed fish have been caught by anglers and another 10,000 by nets. And then there are the offshore English and Irish drift-net fisheries. "There's not one penny that comes from that fishery to support the management of this river," Ronald Campbell says. "Some estimates have said something like 40 per cent of fish that are coming back to Tweed never make it because they are taken up by the drift nets."

There are still six salmon-netting stations on the Tweed — 30 stations have been bought out and retired by anglers. The foundation is using the remaining netsmen in its scientific work, such as radio tagging. "The knowledge they have of the river and the fish is absolutely phenomenal," says Duncan Glen. "Generations have experienced it and we'd be wrong not to pay heed to that."

As we headed back to Drygrange after touring the Tweed watershed, Glen made it clear that the foundation knows it must win over the people of this valley to the conservation cause. "What we're trying to do is win hearts and minds. We're trying to show everybody, not just landowners and farmers, but everybody, that what we're doing is compatible with farming, forestry, tourism, leisure — everything. And it's not always easy. We're working on it."

*

In the final pages of *Pilgrim's Way*, which was published in 1940, three years after Buchan came to Canada, he reflects on how in his lifetime mankind had moved forward rapidly, "perfecting life's material apparatus." This movement, Buchan said, has produced a mood of self-confidence and pride among us. "Our peril has been indifference, and that is a grave peril, for rust will crumble a metal when hammer blows will only harden it."

Today, the world is on the technological fast-track more than ever before. And our peril is still indifference. I spent a day travelling through the Tweed watershed. In late afternoon I climbed up to Scott's View and contemplated a valley filled with history. I went back down into the valley to see Melrose Abbey at sunset. It was a humbling experience. I travelled from New Brunswick to the Tweed, confident that I knew something about salmon rivers. I made my way home filled with more questions than I could answer. As Miramichi river guide Ernest Long says: "Anyone who thinks he knows it all, doesn't know the half of it."

Buchan concludes his pilgrimage with a similar thought. The man who travelled from his child's world in the Tweed valley to Canada's great salmon rivers and back again, ends *Pilgrim's Way* with these words: "Dogmatism gives place to questioning, and questioning in the end to prayer."

Scotland's salmon business

He is attending to the management of the universe.
I am managing my own business.
— Arthur Johnstone

The sun blazed hot over Scotland's thirsty salmon rivers in the summer of 1995. Scotland, a damp country that is used to cool, wet days, suffered through its hottest, driest summer in 500 years. When I was there in August, I was more likely to encounter a sunbather stretched out on the bank of a Scottish river than the tweed-clad angler I had expected to see wading through the country's famous salmon pools.

By the third week in August, Bruce Sandison, the angling columnist for the *Scotsman*, had seen enough. In his Saturday column he wrote a lament for Scotland's rivers, and he launched an attack on river managers and a system which he said had "failed miserably" to care for the Atlantic salmon properly. "Northern rivers have virtually disappeared. Bare bones and dry beds. Flowers grow where salmon should be resting. The story is even sadder in Highland hotels. Not a fish to be seen. Glum faces. Cancelled bookings. Empty bedrooms. The season is almost done and only a bowl of porridge and a lump of peat waits to see us through the winter months. This summer has been a fishing disaster, apart from the netsmen, who prey upon estuary salmon waiting for the rain to help them run their rivers."

Sandison also said that the rivers are suffering from pollution, poorly run forestry operations along their banks and poaching. He blamed riparian owners who have sacrificed environmental probity for profit. "If you balance the books as far as salmon and sea trout are concerned, it does not take an expert accountant to realize that the cupboard is nearly bare and that we have been living on our assets for too long. A few more hot summers might empty the larder completely."

The waters of the River Dee in Scotland were low and hot during the heat wave in the summer of 1995. PHILIP LEE

Scotland's river management system has indeed allowed the country to live off its assets — for the 800 years since the first Scottish salmon conservation legislation was passed in the 12th century. Like other countries with Atlantic salmon rivers, Scotland is fighting to preserve a threatened resource. Poor environmental conditions in the ocean are putting enormous pressure on fragile fish runs, and conditions in the rivers, as Sandison points out, are not as they should be.

In *Tragedy of the Salmon: The Scottish Fishery and the 1986 Salmon Act*, David Shaw, a longtime critic of Scottish river management, harshly criticizes riparian owners for acting in their own self-interest and putting their resource in jeopardy. Shaw's book is a detailed examination of new salmon legislation for Scotland that was passed by Parliament in 1986. Shaw argues that the bill failed to provide necessary controls on Scottish coastal salmon nets, and failed to give the government more central regulatory power over the fishery. "A few very wealthy and often titled people still exercise what amounts to 'local monopolistic' control over virtually the whole of Scotland's wild salmon resource," Shaw wrote in the *Guardian* in May, 1987.

Critics like Shaw and Sandison should not be ignored. Shaw firmly and repeatedly points out that netting wild salmon for food no longer makes sense because these fish are much more valuable in a recreational fishery, and because in some cases the nets could threaten the future of the resource.

Nevertheless, despite the criticisms of Shaw and Sandison, this small, heavily populated and highly industrialized country has succeeded in preserving many of its salmon rivers, some of which are still among the best in the world, when most countries in the North Atlantic have failed. Why does Scotland have world-famous salmon rivers when the United States has none? Why do Scottish rivers still produce almost as many Atlantic salmon as all the Canadian rivers combined?

In light of what has happened to rivers in North America, Scotland's record of salmon management is far from a miserable failure. And today, river managers on some of Scotland's most important rivers — like the Tweed and the Dee — are among the most progressive anywhere in the salmon world.

*

Scotland's 200 salmon rivers are found in three geographic areas: the Southern Uplands, the Central Lowlands and the Northern Highlands. The Southern Uplands are the rolling heather-covered hills that begin on the English border. The famous River Tweed and its many tributaries dominate the eastern part of this region. The Central Lowlands are home to the rivers Doon and Clyde. The Clyde has suffered badly from pollution, but salmon recently returned to the river upstream from Glasgow after a 100-year absence.

The Northern Highlands are home to the country's major salmon rivers (although some of these rivers share highland and lowland geography). The Tay and its tributary, the Tummel, are important salmon rivers. The North and South Esk are the site of a major Scottish river-science program. The Dee is the river of the Royal Family and home of Balmoral Castle, the Queen's summer residence, where the family owns 10 miles of salmon water. Other important Highland rivers include the Don and the Spey. North of Inverness run the Beauly, Conon, Carron, Oykell, Cassley and Shin — all fine salmon waters.

Robert Williamson, Scotland's inspector of salmon and freshwater fisheries, believes his country's system of private river ownership has helped to save the Atlantic salmon resource. PHILIP LEE

On a dusty Edinburgh street lined with government office buildings, far from the glamour of Princes Street and the city's magnificent gardens, mighty castle, and throngs of summer tourists, I met Robert Williamson, Scotland's inspector of salmon and freshwater fisheries. This sharp, friendly man holds a prestigious position that was created in 1882. His predecessors, W.L. Calderwood and W.J.M. Menzies, are legends among salmon fishers in Scotland, and their reports fill his office. Williamson is a student of the history of Scottish river management. He was appointed inspector in 1979 and has been working in the salmon conservation field for more than 40 years.

The first laws to protect Scottish salmon were passed in the late 12th century. Williamson points out that subsequent acts in 1318 and 1457 protected smolt on their way down rivers into the sea. Legislation in 1863 had a threefold purpose: to ensure that salmon had access to the upper spawning grounds; to protect the smolt run to the sea; and to prohibit the killing of fish at spawning time.

The foundation of Scotland's management system has always been property rights. "The unique element in Scotland, and it has always been, is that salmon fisheries in the sea as well as in the rivers are privately owned. These private and marketable rights are the basis of salmon-fisheries management in Scotland," Williamson says. Salmon fishing rights can be bought, sold or leased apart from the land. Scotland's private salmon fisheries in the sea are connected to specific areas of coastline and extend to the 12-mile limit.

Scotland's salmon legislation divides the country into more than 100 salmon fishery districts. The owners of the salmon fisheries may set up a district fishery board, which is a self-financing body that works to protect and enhance the rivers. The district fishery boards have the authority to tax the owners of salmon waters, but the boards do not have the power to make specific fishing regulations. These must be applied for through Scotland's Secretary of State. The owners elect members of the district fishery boards: three members from the upper stretches, three members from below the tide line and a chairman.

The permission of the owner of the water rather than a government licence is required in order to fish salmon in Scotland. Owners can fish as much or as little as they want. Williamson says the government has outlawed methods that might destroy the stocks. In a river it is legal to use rod and line (some but not all rivers are open to fly fishing only), or net and coble (a small boat), a method of harvesting salmon in which the net makes a circular sweep through the river but can't stay in one place and obstruct fish (like the gill nets in New Brunswick rivers).

In Scottish ocean waters, drift nets and gill nets are banned. The only accepted methods of fishing are stake nets and bag nets, which trap salmon as they swim along the coastline searching for the outflow of their home river. A stake net traps salmon in nets held up by stakes in the sand. A bag net floats and is used in deeper water.

All fisheries must be closed for 168 days a year. Angling is closed on Sundays and net fishing is closed from 6 p.m. Friday until 6 a.m. the following Monday. Recreational angling is big business in Scotland. The cost of a day's fishing ranges from $25 to $2,500. Good salmon waters are extremely expensive, selling for as much as $35,000 for every salmon caught in the waters in a year. A fishing beat that produces an average of 100 salmon a year would sell for $3.5-million, more than twice the price for a similar beat in New Brunswick. Some private owners set catch limits, but limits are entirely up to them. Fish caught on a rod are the property of the angler and can be sold.

Catch-and-release fishing, the cornerstone of North American fish management, is seldom practised in Scotland. "I think it's potentially a very dangerous idea," Williamson argues. "Going to a river to fish with the idea that every fish you catch is going to be put back, and that this is somehow more noble than catching a fish and killing it and eating it, is to my mind sophisticated nonsense and, in a practical sense, it is more dangerous than just sophisticated nonsense. Because, sooner or later, somebody is going to say it's not morally tolerable for people to go out for the fun of hooking a fish and playing it and turning it loose, having sport at the expense of that animal."

*

While there is tension on the rivers between anglers and netsmen, Williamson insists that Scottish netsmen can't be blamed for declining fish stocks. "If, in the larger Scottish rivers, there is not enough salmon to allow both an economically viable net fishery and good angling, then there is almost certainly something more seriously wrong with the salmon stock of the river than just the operation of the local nets."

There is no place, however, in Scotland's waters for a common-property net fishery. When fishermen began drift-netting for salmon off Scotland's coast in 1960, the government took immediate steps to stop them and banned this new free-for-all fishery by the end of the 1962 season. "Scotland was the first major salmon-fishing country to prohibit the use of drift nets. It was done on a properly argued scientific basis before any damage was obvious in the fishery — a far-sighted decision and fortunate for the well-being of Scottish salmon stocks and fisheries," Williamson says. Scottish salmon fishery owners pushed for the ban to be extended to the northeast England drift-net fishery, but they failed. England and Ireland are still drift-netting for salmon off their coasts, killing fish destined for rivers all over the United Kingdom. The fact that there were private fisheries operating up and down Scotland's coast allowed the government to act decisively against the gill nets, Williamson says.

"It was a profitable fishery. The catch was going up. There was no indication that it was about to collapse, yet the government bans drift-netting. I think that would have been politically quite difficult to do if all the fishing interests had just been the participants in a public-ownership fishery. The fact that the fisheries within three miles at that time were all privately owned meant that all those with an interest in the long term could accept that this was not a sensible long-term project and we should ban it. I think that's why in Scotland it was banned while in England it wasn't."

*

Williamson says the coastal net fisheries adjust themselves to the number of salmon returning to the rivers. If there is a great abundance of salmon, then the netsmen use more nets. If there are fewer fish, then they use fewer nets. The netsmen and the anglers are also balancing themselves out. With the development of farmed salmon, the value of the net fishery has gone down. Meanwhile, the value of good salmon angling continues to rise. "It is quite open for those who own the rod fisheries in the river to buy the net fishings and operate them at half the intensity, or not operate them at all," says Williamson. On the River Dee in the late 1800s, the salmon proprietors upstream bought out all the nets below the town of Banchory, operated them for a few years until they could sink their debt, and then shut them down. On the Tay in the late 1890s, the upper river owners set up a company to buy out nets in the mouth of the river. They operated half of them. The value of both the rod fishings and the net fishings increased. More recently, a Scottish group called the Atlantic Salmon Conservation Trust has been buying out coastal netting rights as a contribution to the conservation of the resource.

Each river system is different, Williamson says, and this flexible structure allows for changes in economics, social behaviour and fish stocks. For more than a decade there has been a fierce debate in Scotland about whether the government should have the power to act decisively on rivers in the interest of conservation. "There is an advantage to be able to act quickly if there is a disaster, but there is an inherent risk in having powers to act immediately," Williamson says. The government needs to strike a balance between having the power to make instant changes and having the sense to wait and see whether the situation works itself out. "I think one of the advantages of the system we have is that it doesn't require direct, day-to-day, hands-on management by a bureaucratic organization."

*

In almost every respect, David Shaw and Robert Williamson part company. Shaw argues that netting interests have dominated local fishery boards and the future of the resource is threatened. His book concludes that Parliament had a chance to shift the balance in 1986 but failed. "Netsmen, who continued to take up to 97 per cent of the summer catch while accounting for less than 5 per cent of the benefits generated by the Scottish salmon economy, preferred, understandably, to operate with minimum restraints," Shaw writes. "Nor were dominant rod proprietors prepared seriously to challenge this staggering imbalance. Most were reluctant to support reform initiatives while still they found themselves in possession of fisheries which, despite heavily diminished catches, continued to command high fees and showed steady increases in capital values. Futhermore, if rod fishery owners pushed too hard for statutory restraints on downstream netsmen, what might be the reciprocal restraints they would have to accept? 'Conservation' was acceptable only so long as it might result in larger catches, as interpreted by net or rod interest. In any other sense 'conservation' was at best regarded as distasteful, at worst an extremely suspect, even dangerous concept."

The debate over the 1986 act underscored the opposition between the netsmen and the salmon anglers and conservationists who wanted the nets to be curbed by law. Shaw maintains angling and tourism generate more than $400 million annually and create 30,000 jobs, compared to the netsmen's haul of $10 million and 800 jobs.

The debate in Parliament and the House of Lords centred philosophically on the question of whether private property rights should be subject to central regulatory control. Writing in the *Glasgow Herald* on May 21, 1986, during the debates on the new salmon act, Shaw said:

> Few would dispute that the Highlands Region is perhaps the leading anachronistic social curiosity of the modern Western world. It is a region where a residual aristocracy has managed through its hold on often vast properties to retain an extraordinary level of influence and often community authority. The practical result is that what these eminent persons wish to be legislated will often be legislated. More to the point, what they do not wish to be legislated, won't. As many of these persons retain important, and not so important, netting rights they are steadfastly opposed to any change in the salmon laws which might result in the smallest encroachment on these rights. Basically, as perhaps most clearly demonstrated by Lords Lansdowne and Thurso in the Lords Debates, these elements are determined to retain the privilege of excercising their rights in almost exactly the form they have for the past 100 years. Given this position is it then so strange that the current Bill has been described as the "archaic body of the 1862 Salmon Fishery Acts in new clothes"?

The 1986 act re-emphasized the private rights of the fishery owners, tightened up on illegal fishing in the ocean, and increased sanctions against poachers in rivers. Shaw notes that Lord Lansdowne, one of the most powerful salmon fisheries owners in Scotland, dismissed the new bill as a "wee tim'rous beastie." Shaw says that it was "a public Bill aimed primarily at promoting and protecting private interests — a Laird's Bill." Scottish Labour MP Norman Buchan said: "We know what the salmon means in Scotland, and especially in the Highlands. The salmon means privilege. That is what it is all about in the Highlands, where King Salmon still rules." The Opposition, which wanted greater public access and control, and more central regulatory authority, called the bill a Landlord's Charter.

Shaw concedes that there were some positive measures brought forward in the new act. The act allowed fisheries board chairmen to be elected by all proprietors. "Certainly, the first steps, albeit few, limited and fragile, had been taken towards reconstruction of local fishery management in Scotland." Under 1860s legislation it was unclear whether boards were private or public statutory bodies. The 1986 act made it clear that they were private.

"This then was the new Salmon Act which had been produced," Shaw writes. "It failed to meet established century-old requirements. It set out local management provisions which were regressive even in 1860s terms. It did, however, provide modest hopes of improved protection. And despite its alarmingly antiquated provisions relating to exploitation, it was neverthless possible that improvements to meet modern conservation, economic and equity aims might at least partially be attained." Shaw says it is possible that farmed salmon may make netting of wild salmon economically unsound. And he noted that the Atlantic Salmon Conservation Trust may buy up more nets. While still a scathing critic, Shaw remains hopeful for the future.

✻

Shaw ends his *Tragedy of the Salmon* by describing a scene from the spring of 1990 when he visited a salmon restoration project at Wilmot's Creek, a former salmon river that flows into Lake Ontario at the town of Newcastle. Shaw's guide, Ontario Natural Resources biologist Les Stanfield, pointed out that nets in the early 19th century took 50,000 fish a year from the creek and drove the salmon runs to extinction. In recent years, biologists have introduced steelhead and Pacific salmon into the creek to try to create a sport fishery.

Shaw took great comfort in the fact that there would be no more net fishing at Wilmot's Creek. "As we drove back to Toronto, I reflected on another Newcastle I knew, a Newcastle facing out onto the North Sea. I thought of the drift nets, the 'self-policing' fishermen, the coastal fixed nets, the intensity of commercial fishing, with — unbelievably — the prospect of worse to come." On the other hand, he says the Ontario experience gave him inspiration, support and encouragement.

For all of Shaw's good intentions and admirable passion, his argument is flawed. The Wilmot Creek salmon are extinct, like salmon in many North American rivers. Atlantic salmon in Scotland are in trouble in some rivers, but they are not extinct. Scotland may have to change its ways and stop killing so many fish, but it still has something left to save. Shaw won't discover the way to the salvation of the Atlantic salmon in our empty waters.

The royal river is losing its crown

I desperately want to see the salmon survive as a species — not just because I want to catch them, but because their disappearance would be a tragedy of major proportions and would matter to us all.

— Prince Charles

The Royal River, Scotland's Aberdeenshire Dee, is in trouble. Lieutenant Colonel Robert Campbell, the chairman of the Dee District Salmon Fishery Board, is charged with saving the famous river, and the magnitude of this responsibility is not lost on him. An energetic retired Royal Marine, his estate overlooks his private fishing beat on the Dee, about an hour's drive downstream from Balmoral Castle, the Queen's summer home.

These days, Campbell rarely finds the time to cast a line in any of his 11 private pools. The famous spring Atlantic salmon runs on the Dee are declining rapidly, and the fishery board is fighting the conservation campaign of its life. The day I met Campbell, on a summer afternoon in 1995, he was fielding calls from the local press on his cellular telephone and at the same time preparing an application to the Secretary of State to legislate controversial new conservation measures for his river.

"We've got a serious problem of a shortage of fish," he told the reporter from a

The 200-year-old Banchory Lodge is one of the famous River Dee fishing hotels downstream from Balmoral Castle. PHILIP LEE

local newspaper. "The district fisheries board must act if there's evidence things are bad. Either we go for conservation measures or we don't." What about a report produced by a riparian owner who opposed the conservation measures and said Campbell was going too far? the reporter asked. "There's an awful lot of woolly words in it," Campbell replied with a smile and a wink for his Canadian visitor.

Salmon run up Scottish rivers every month of the year, but spring salmon, which come into the rivers before the end of April, are the largest and most highly prized fish in Scotland. Spring salmon on the Dee spend more than one winter at sea, mingling with North American salmon on the ocean feeding grounds before returning to the river between November and May. The average spring salmon on the Dee weighs about eight pounds, although some stay at sea for several winters and grow much larger. But these spring salmon are dying at sea and no one knows why. The Dee is facing the same problems as New Brunswick salmon rivers; this is the story of the Dee and Campbell's fight to save it.

*

To get to the Dee and Campbell's estate, I drove north from Edinburgh to Perth and then across the Cairnwell Pass, the highest road in Britain. This barren landscape is spectacular in late summer, with round hills covered with purple heather in full bloom. The tiny Clunie Burn rises on the left side of the pass, tumbles down from the mountains and joins the Dee at Braemar, the site of Braemar Castle and a cottage where Robert Louis Stevenson wrote a good portion of *Treasure Island*.

The Dee, the highest-altitude salmon river in the United Kingdom, rises from the Well of Dee in the Cairngorm Mountains, 4,000 feet above sea level. At Braemar, the cold, clear and unpolluted river runs into a narrow, wooded valley. On a right-hand turn in the river below Braemar sits the majestic Balmoral Castle, and from there the Dee continues to widen down through the angling town of Banchory and on to the city of Aberdeen, where it runs into the sea. Salmon fishing on the Dee is comfortable — no driving down dirt roads, tramping through the woods and getting swarmed by insects here. At the 200-year-old Banchory Lodge, one of the famous Dee fishing hotels, guests stay in elegant rooms, eat gourmet meals, drink fine liquor and stroll across a wide lawn to cast into a most scenic salmon pool.

Opposite the Banchory Lodge, the cold Water of Feugh flows into the hotel's salmon pool. The Falls of Feugh attract thousands of visitors each year to watch the salmon leaping up the rapids. Dugald Jaffray, the owner of the Banchory Lodge, charges his guests about $150 a day per rod for fishing. He remembers when they were landing 30 fish a day at the hotel. "Over the years there's certainly been a decline," he says.

Jaffray has seen both Prince Charles and the Queen Mother land salmon on the Dee. The Queen Mother, he says, used to be a "marvellous caster," and Prince Charles is teaching both his boys to fish. Howell Raines, who worked as a *New York Times* reporter in England, writes in *Fly Fishing Through the Midlife Crisis* that "killing a salmon is one of the ways by which Britons determine social rank and those within the higher ranks prove their importance to themselves."

Lieutenant Colonel Robert Campbell, seen here in one of his private pools, is leading the conservation movement on the River Dee.

Courtesy of Col. Robert Campbell

"As evidence, consider that one of the few honeymoon pictures of the Prince and Princess of Wales released by Buckingham Palace shows Diana posing in front of four large dead salmon. The 1981 photograph was taken on the Scottish estate of the Duke of Westminster, the wealthiest landlord in London. Diana, clad in country clothes, stands casually, one lanky leg crossed over the other, like a dancer at rest. She has linked her arm through that of her gillie, a stout man in a tweed suit, matching cap and dark tie. As the Princess leans against him, the grey-haired man beams with fatherly pride. In a society attuned to interpreting the signals of rank, the intended message from the Palace was clear. Here is the queen-to-be, able to reap fish from the streams of the wealthiest and command the service and adoration of the hoi polloi." Word along the Dee has it that despite this honeymoon photograph, Diana never really liked to fish for salmon.

The Royal Family owns five salmon beats at Balmoral and controls fishing on several important tributaries, including the Girnock Burn, which the Queen leases from her next-door neighbour. This all adds up to about 12 miles of prime salmon fishing. Robert Campbell says: "It's a lot of fishing. Prince Charles is a great enthusiast and he's taken a great interest in what we've been doing and has given us a lot of support and a lot of encouragement. He's sent for me and another member of the board and a member of the improvement association and quizzed us and asked what help he can give."

Speaking to an international salmon symposium in 1978, Prince Charles foreshadowed what would happen on his own river a decade later: "Pressures on any

Rear Admiral John MacKenzie is director of the Atlantic Salmon Trust, a conservation and research organization in Scotland. PHILIP LEE

resource nowadays are becoming almost intolerable as a result of man's curiously insatiable greed and his equally curious and apparent inability to understand the necessity for sensible conservation and wise management. 'There is no *proof* that the stocks are being depleted,' seems to be the perennial cry. It is only when the catches are reduced to almost nothing that self-interest stimulates some kind of action — and the awful thing is that it could be too late."

Campbell hopes he is acting before it is too late. "We are probably the only river in the world with almost entirely spring fish in it, and it's the spring fish that have disappeared. We know that the high mortality rate is in the sea. The numbers of adults returning per smolt going to sea is down from about 40 per cent to about 5 per cent of spring fish. Your percentage of fish returning is going down, down, down."

Since the late 1950s, rod catches in the Dee have dropped from 12,000 salmon a year to 3,500. Net catches have dropped from 25,000 salmon to about 8,000 in the same period. On Campbell's beat, fishermen used to average 270 spring salmon by the end of May. Now, they average about 37. What is happening to the Dee's springers?

*

At the office of the Atlantic Salmon Trust, the wild salmon crises in Scotland and New Brunswick find common ground. The trust, based in the Highlands town of Pitlochry, is one of Scotland's leading wild salmon conservation organizations, and its overriding concern these days is the mysterious disappearance of large salmon from Scottish rivers. The decline in these springers began in the 1960s but has accelerated in recent years. The mostly female springers carry large numbers of eggs and are a unique genetic strain, so when they start to disappear, they disappear quickly.

In 1995, in typical scientific jargon, the International Council for the Exploration of the Sea reported on the state of spring salmon: "For stocks in which spawners were consistently below replacement, spawning stocks generally declined toward the origin. This tendency should be viewed with alarm because it reveals the loss of a run component and, possibly, any genetic attributes that may be associated with a high sea age." Rear Admiral John Mackenzie, the Atlantic Salmon Trust's director, translated this statement into plain words: "This means that stocks are approaching zero."

"There are problems at sea," he said in an interview as we drove south from Pitlochry to visit a fish hatchery in Perth. "The most difficult thing in the world — and the most expensive thing — is to try to do any research into the salmon at sea because, surprising though it may seem, the salmon in the sea is a very rare fish. And it is very difficult to actually find out why we're getting this very high mortality at sea."

The Atlantic Salmon Trust has developed a series of specific policies to deal with the alarming decline in spring fish: an end to industrial fishing of sand eels and capelin, important food sources for salmon in the sea; an end to all fishing of mixed stocks of salmon at sea; the permanent closure of drift-net salmon fisheries off the English, Welsh and Irish coasts; the delay of the coastal salmon netting season until the spring fish get into the rivers; the restriction of angling; and, in the long term, the cessation of all salmon netting.

❋

A river will produce only a certain number of smolt, and when a river has enough spawning salmon to produce this number year after year, the river is as well-managed as it can be. Scientists are working on the Dee to find out exactly how many spawning fish the river needs, both on the main stem and in every tributary. Each tributary, Robert Campbell says, has its own distinct genetic stock of Dee fish.

If you lose that magic number of spawning fish, "you fall off terribly quickly," Campbell says. "Your rivers start producing half a smolt run. You can be at the point of no return very, very quickly." The Dee is a high-altitude river with a special type of salmon that grows slowly and lives three or four years in fresh water before going to sea. After several years of low populations of young salmon leaving the river, smolt start going to sea too early to try to compensate. When this happens, the fish stocks of future years are compromised and the whole process of decline speeds up.

Scientists have been doing extensive research on the Girnock Burn, the tributary that enters the Dee near Balmoral Castle. The Girnock has only about half the young salmon it needs. Despite the fact that Scottish fisheries like to be left alone to manage their own waters, the fishery board had to take action, Campbell says. If he did nothing, his "head would roll, and quite rightfully so."

The fishery board realized that it had to stop the number of fish being killed on the river, and so it set these rules for conservation: The opening of the season for

A stained glass window at the Banchory Lodge: After clouds, sunshine. PHILIP LEE

rods and nets was to be delayed by one month, from February 1 until March 1; anglers were to use flies only; anglers were to limit themselves to one spring fish per week; all three-sea-winter fish were to be released unharmed.

Campbell's measures were not greeted with enthusiasm by all fisheries owners. And at the same time, the Dee's new electronic fish counter let him down, missing a large run of fish that passed over it during high spring water flows. Some riparian owners began to question whether the Dee was in trouble at all. Dugald Jaffray at the Banchory Lodge says he has no problem with conserving the resource and limiting his guests to one fish per week; he just doesn't want to be told that he has to do it. But Robert Campbell says that asking people to put fish back in the river is simply doing what Americans and Canadians have been doing for years.

Despite opposition, the conservation campaign seems to be working. In 1994, 82 per cent of large salmon caught on the Dee were released and 60 per cent of all fish were put back. "There's been a sea change of attitude in this country and I'd like to think we have played a significant part in it," Campbell says. "We've got to take vigorous steps — firm steps for conservation — before it's too late."

❋

Campbell is following in a long tradition of active salmon conservation on the Dee. In 1872, the Marquis of Huntley, who owned 30 miles of fishing near the top of the river, established the Dee Salmon Fishing Improvement Association with the goal of buying and renting salmon net fishings below Banchory and removing the nets.

"The most important single factor in making this river into such a successful salmon fishery was the foundation of the Dee Salmon Fishing Improvement Association," write Derek Mills and Neil Graesser in *The Salmon Rivers of Scotland*. The buying of nets continued gradually over the years as they came on the market. Nets in the Aberdeen Harbour were bought out in 1986. All of this was done with private money contributed by the people who owned the fishing rights in the river.

There are still coastal nets near the mouth of the Dee. When I was there the Dee association was negotiating a purchase of a netting station at a cost of about $50 a fish, based on average catch per year. So a netting station bringing in just 1,000 fish a year would cost $50,000.

Some of Campbell's critics say the reason for the decline in Dee springers is the board's decision in the late 1980s to close its hatchery, which needed a $200,000 refit. Scientists came to the board and said that there were better ways to spend the money than on a hatchery. "The hatchery was almost certainly doing very little if any good," says Campbell.

To answer his critics, he looked up the rod-catch statistics from the five years before and after the hatchery was put in. With the hatchery operating, "there was a 36 per cent decline in spring fish, an 18 per cent decline in summer salmon and a 13 per cent increase in grilse." Hatcheries are good public relations, but that's about all, Campbell argues. The fishery board and improvement committee are spending their money on eight bailiffs to stop poaching and one biologist to lead a research program.

The fishery owners pay a portion of their profits to run the river. The Dee fishery board needs to raise about $650,000 a year to pay for its operations. Its problem now is how to raise money at a time of seriously declining catches. The nets are catching more than 50 per cent of the fish but paying a very small part of the expenses because they are not making money. As a result the anglers are footing 98 per cent of the bill.

Campbell says this dilemma reveals a fatal flaw in the Scottish system, and the only answer is a reallocation of the resource. Salmon farms are producing huge numbers of fish, so why net wild salmon? "Commercial netting for salmon is obsolete — the equation just doesn't work. They cannot pay for the resource with their profits, so the only way that they could pay more is to catch more fish — and there aren't more fish to catch."

Campbell was elected to another term as chairman of the Dee District Fishery Board in August, 1995. These will be critical years for the future of the Dee. A stained glass window at the Banchory Lodge features a leaping Dee spring salmon and words of hope for a river in troubled times: *Post Nublia Phoebus*. After clouds comes sunshine.

Netting the North Esk

Salmon nets have been a fixture for more than 150 years on the beach of Montrose Bay, which runs for five miles between the North and South Esk rivers on Scotland's east coast. For many years, the owners of the netting stations and the angling beats have played an active role in salmon conservation and scientific research on the river. As a result, despite the fishing pressure, the salmon runs on the North and South Esk have remained relatively strong.

As I approached Montrose Bay, the sun and wind pushed the fog from the white sand beach and the salmon nets appeared. Fishermen, who looked more like circus performers in a balancing act, walked along ropes running 25 feet above the sand across the tops of the stake nets. These nets stretch out from the beach on stakes and trap salmon as the fish swim along the coast, searching for the mouth of their home river. Other fishermen were preparing to go out into the waves in a small boat to harvest salmon from floating bag nets; moored out in deep water, these nets also capture salmon as they return to their home waters.

Fisherman Brian Reid backed a wooden boat down the beach into the sea. Ed Dryden manned the oars while Euan Grant sat with me and waited for the action to start. After rowing out of the shallows, we motored to the net stations where the three of them pulled the heavy nets out of the water and dropped live Atlantic salmon into the boat. Grant killed them by hitting them on the head with a club. Reid, Dryden and Grant emptied four bag nets in about an hour and brought in 11 large salmon. When we returned to the beach, they packed the salmon in wooden boxes and covered them with ice.

The fishing at Montrose has been steady but slow. "They come in cycles," notes Bob Richie, who has been working the nets for 35 years. He said the bag

A salmon netsman on the beach in Montrose Bay, Scotland, looks more like a circus performer than a fisherman. PHILIP LEE

and stake nets are not efficient: If 20 salmon are trapped in the nets before dark, 15 of them will escape overnight. But this is the only type of gear they are allowed to use; therefore, the fish are safe from being destroyed by improved technology, one of the major threats to the long-term health of stocks in most ocean fisheries. "We only take part of the crop," Richie said. "It ensures a good number of the fish go into the rivers."

Netsmen clear a salmon from the mesh of a bag net in Montrose Bay. PHILIP LEE

✻

The nets on the beach are the property of Joseph Johnston and Sons, a salmon-fishing company that also owns netting and angling rights in the river. It has been involved in the salmon-fishing industry in Montrose, a working-class Scottish seaport city, since 1826. In the company's office on the old Montrose waterfront, I met the chairman, Noel Smart.

Smart, a sharp, lean, intense man, has followed in the footsteps of his father, George Graham Smart, who ran the company before him. He works out of an old, Spartan office, and sits behind the same giant wooden desk where his father sat to manage the company's affairs. Joseph Johnston and Sons used to have netting stations in other parts of Scotland — on the Isle of Skye, the Fyfe coast and the Aberdeenshire coast — but now its operations are confined to the area around the North and South Esk.

The price of Atlantic salmon has dropped since the market became flooded with farmed fish, but top-quality wild salmon are still worth more than cage-raised fish. The company operates sweep nets in the estuaries of the North and South Esk, fixed nets in Montrose Bay and one netting station three miles south of Montrose. The company also sells rod fishing on two-and-a-half miles of the North Esk.

One of the major criticisms of these coastal netting stations is that they take fish that are migrating along the coast to a variety of Scottish rivers. Author David Shaw calls the Montrose nets the country's leading interceptory fishery. "For its size, it was probably the world's most productive Atlantic salmon commercial fishery," he writes.

However, according to Smart, the future of Joseph Johnston and Sons is tied to preservation of the salmon runs in these rivers. "The company has always had a policy of managing its affairs to suit salmon stocks. We're a company that's always had a commercial interest in salmon netting. We've always believed that when you manage a river properly there's plenty of room for a commercial netting operation and a successful rod-angling operation at the same time. There's plenty for both, but I think it's management of the river system that is the contributing influence in that."

Joseph Johnston and Sons has financed its own river-enhancement projects. For example, in the gorge on the upper North Esk there is a series of waterfalls called The Loups of The Burn. (The Burn is an 18th-century mansion and estate.) One of the falls had been impassable for salmon in low-water conditions. In 1948, Smart says, "The company actually made a natural fish ladder out of rock there, which opened up a whole tributary and had a very marked effect on the numbers of fish that went up there, and as such they spawned and increased the returning stocks."

A branch of Scotland's government freshwater fisheries research laboratory operates a field research station in Montrose and has done extensive scientific work on the North Esk. Scientists are measuring the number of fish the river is capable of producing and carefully counting the number of fish that return. The North Esk was chosen because it is a fairly short river, about 30 miles long. It is also a productive salmon stream, and the established netting operation at the mouth has allowed scientists to build up data on the fish returning to the river.

Smart thinks the river is "the most researched river in the world." An electronic fish counter installed 21 years ago recorded 11,000 fish in 1994, the highest registered upstream count of fish to date. "The numbers of salmon going upstream — even allowing for some to be caught by angling and even poached — is many, many times more than what is required to see the survival of the stock here in the North Esk. It would be a multiple of tens of times the number of fish required to see the survival of the stock in the North Esk."

The research station also has a facility for trapping smolt coming downriver in order to get a population estimate: "Your river is the basis for everything that happens. If you don't produce smolt in your river, you can't expect to get fish back. Equally, there is a maximum production of smolt in any given watershed, depending on a number of factors. It doesn't matter how many eggs you put down there, you will not get above that maximum. So as long as we can ensure that all the river is populated with small fish in juvenile stages, then there is no more we can do in the river to make sure more fish come back. It's a self-limiting process. So we have these smolt estimate counts on the North Esk and there is no discernable trend, neither up nor down, over the past 20-odd years. So it's clear that the river is producing the maximum number of smolt to go to sea.

On the other hand, Smart observes, "they are not returning in the same numbers as they used to. And there are many theories as to why this should be, and nobody can actually pin anything down on that. But there's nothing more you can do within the river system to produce more smolt, apart from rehabilitating some spawning areas and some juvenile areas, which may or may not have been eroded and become debilitated over the years."

The local fisheries board is rehabilitating a tributary of the North Esk called the Mark, which scientists have discovered is a source of spring fish. Workers hired by the fishery board are stabilizing its banks, and, on a slow netting day in the summer of 1995, netsmen from Joseph Johnston and Sons were planting trees and fencing off the river to prevent the sheep from grazing down to the water. The week I was there, they were conducting surveys of juvenile fish and installing a fish counter.

The size of the run of young salmon into the sea changes from year to year. For example, the summers of 1994 and 1995 were very dry and the fish were not able to penetrate some of the small tributaries. "We always have a reserve," Smart says. "We take some brood stock in the autumn out of the river, take the eggs off them and put them in a hatchery. Where we feel there isn't enough natural spawning, we will put fish from natural North Esk stock into those side burns, which we did this year. So we try to make sure that there is seed scattered in all the feeding areas." One river tributary is blocked by impassable falls where it is not economically feasible to build a fish ladder; however, there is a marvellous natural juvenile feeding area above the falls. Therefore, every year workers stock that area above the falls and the smolt have no problems coming down over the falls into a deep salmon pool.

*

However, the question still remains — what is happening at sea? Smart thinks "pollution, in the widest sense, is causing the problem." Discharge from detergents, disinfectants, bleach and cleaning agents is being washed into the sea. This discharge, he thinks, may be causing the death of lower forms of life in the food chain along the coastline. "That is the basis of the food chain and I think there is no doubt that there is less small fish life and bug life along the coast than there used to be. The bottom of the food chain is missing. These smolt and these sea trout which do feed locally before going off will be in big trouble. And perhaps that's the reason why the whitefish in the North Sea seem to do so poorly as well." (Whitefish feed and spawn close to the shoreline.) What is happening further out in the North Atlantic, far from the netting stations on the beach in Montrose Bay, remains a mystery.

Salmon conservation's United Nations

The United Nations of the Atlantic salmon world works out of an elegant old townhouse on a corner of Rutland Square in downtown Edinburgh, Scotland, a short walk from the magnificent castle that dominates the city's skyline. This is the office of the North Atlantic Salmon Conservation Organization, where international conservation battles will be won or lost in the next decade, and this organization will play a crucial role in deciding whether salmon stocks recover or fade into extinction.

Dr. Malcolm Windsor, the organization's secretary, has been working to ensure a future for the salmon since 1985, when the North Atlantic countries signed a treaty to work to conserve the resource. Windsor says the international community is finally starting to pull together to save the salmon. "I think for the first time there is sufficient concern in the domestic and international community to start turning this around. There is now a mechanism to co-operate to conserve the species. It is beginning to work and I would take that as a positive sign. Of course, it might be too late. Maybe we're seeing the final decline of the salmon. Maybe the stocks 10,000 years ago were so huge and now we're down to a remnant population. Maybe we're just pushing them to the edge, and once we start releasing huge numbers of farmed fish the way we are doing now and the ocean becomes more polluted, that's the end of it. But we can't know that. We can't assume that's happening. All we can do is say that for the first time there is a political will. Every country in the North Atlantic has signed this treaty and they are working together. We have a better chance than we had 10 years ago. As the threats are coming forward, we are dealing with them one by one."

Dr. Malcolm Windsor, the secretary of the North Atlantic Salmon Conservation Organization, outside his office in Edinburgh, Scotland.
PHILIP LEE

Windsor makes it clear that while there are many conservation battles to be fought on the international stage, you won't find any scapegoats in his office. "We have to face up to the fact that we have to change our ways ourselves. We have to look inwards. We have to stop using cars so much. We have to stop damming rivers. We have to stop allowing pollution."

There is a lot at stake here. The salmon industry in the North Atlantic is worth an estimated

$3.5-billion (U.S.) The salmon resource in Great Britain alone is worth more than $800-million (U.S.). But the Atlantic salmon is worth more to people than its market value. “People simply feel a bit happier knowing that salmon are in their rivers and going about their migrations,” Windsor says. “The salmon is a symbolic creature. So if the salmon is in a river near you and you know it’s there, it shows you the water is good and it shows you all is well with the world.”

*

The seeds of the North Atlantic Salmon Conservation Organization were planted in 1978 at an international Atlantic Salmon Symposium in Edinburgh sponsored by Scotland’s Atlantic Salmon Trust and the New Brunswick-based Atlantic Salmon Federation. Conservationists knew the salmon was being heavily fished on the high seas; because fish from North America and Europe were being wiped out as they mingled in distant feeding grounds, the Edinburgh symposium called for an international treaty that would ban salmon fishing in the ocean beyond 12 miles from each country’s coastline. The proposed treaty would encourage all countries to co-operate in salmon conservation, regulation and enforcement, and to develop and share scientific information.

The next year, the U.S. State Department developed a draft treaty, and for the next two years diplomatic meetings were held in Washington, Brussels, Ottawa and Oslo. Despite an alarming drop in the world’s salmon stocks, the international diplomatic wheels turned slowly. Between March and October of 1983, a Convention for the Conservation of Salmon in the North Atlantic Ocean was open for signatures in Reykjavik, Iceland. Canada, Denmark (in respect of the Faroe Islands and Greenland), the European Community, Finland, Iceland, Norway, Russia, Sweden and the United States all signed the convention. Under the convention, the North Atlantic Salmon Conservation Organization was formed to contribute “to the conservation, restoration, enhancement and rational management of salmon stocks in the North Atlantic Ocean.”

Malcolm Windsor was chosen to be the secretary, and the office opened in February, 1985, seven years after the Edinburgh symposium. Windsor works with Dr. Peter Hutchinson, his assistant, and a support staff of two secretaries. Decisions on regulatory measures — for example, how many tonnes of salmon are to be caught off West Greenland — are made at meetings of the organization’s three commissions: the North American Commission, the Northeast Atlantic Commission and the West Greenland Commission. These decisions require unanimous agreement.

*

During its first five years, the organization was preoccupied with bringing some order to the Atlantic salmon ocean fishery, but there was one immediate victory when the convention was signed. In most areas, salmon fishing beyond 12 miles from the coast was prohibited, although there were concessions made on this limit for Greenland and the Faroe Islands. This measure immediately closed the salmon fishery in international waters, including that in the northern Norwegian Sea, where ships from Denmark, the Faroes, Finland, Germany, Norway and Sweden took 960 tonnes of salmon a year at its peak.

The big question facing the organization was how to control the fisheries not banned by the treaty that were intercepting salmon headed for other countries. Canada was intercepting United States fish. Norway was intercepting Russian fish. Greenland and the Faroe Islands were intercepting everybody's fish. "The whole North Atlantic was a pattern of interception," Windsor says. The organization has concentrated on this question, and it has made substantial progress. The Canadian interceptory fishery has been drastically reduced in the last 10 years, and Ottawa has bought out commercial salmon licences at a cost of more than $70-million. (The Labrador commercial fishery continues to take U.S. and Canadian fish.)

The 1995 quota set by the North Atlantic Salmon Conservation Organization for the Greenland fishery, which was taking 2,700 tonnes of salmon at its peak, was 77 tonnes. The 1995 quota for the Faroe Islands, whose catches had peaked at 1,000 tonnes in 1981, was 550 tonnes. (Since 1991, the Faroese fishermen have not caught salmon, choosing instead to accept compensation from the North Atlantic Salmon Fund, a private conservation organization led by Icelander Orri Vigfusson.) Norway banned drift-netting within its 12-mile zone in 1990, but England and Ireland are still drift-netting salmon off their coasts.

This process forced these countries to take a long, hard look in the mirror. "If Norway was going to come and demand that the Faroese reduce their catches, then the Faroese were going to say, 'Well, okay, we understand your position, but you are polluting your rivers, you are allowing drift-netting within your 12-mile zone, you must put your own house in order and then we'll talk,'" Windsor says. The Greenlanders were under enormous international pressure to cut back. "But they said, 'How can you expect us to make all the sacrifices in a country which has no other opportunity to earn a living other than fishing, a country which has no other resources?'"

*

In the early years, the organization sorely lacked solid scientific advice. "It was like driving a car with no windscreen," Windsor says. "It only had a rear view mirror.

The scientists' report every year told us what happened last year and in the previous year. It did not tell us what the abundance might be in future years. If you don't have any advice, which we didn't have for years, how do you decide? Nobody can say it's wrong. Nobody knows. The lack of scientific advice should not be used as an excuse to take risks. You should assume the worst, and only act in that way."

In 1992, international scientists came up with a forecast of salmon abundance, looking particularly at stocks off Greenland, examining the sea surface temperature, the size of the feeding ground and the number of smolt going out to sea. This meant that the organization's quotas were based on scientific research rather than on purely political pressures. In 1995, the organization used these formulas to come up with a 77-tonne quota for Greenland. The organization arrived at this controversial decision by taking the scientific advice, which recommended a complete closure of the Greenland fishery, and applying social and economic factors. The result was a small quota to keep the Greenland fishery alive.

Dr. Wilfred Carter, a man with many years of experience in international salmon conservation negotiations, was deeply disappointed by this decision. "After years of fruitless discussion on methodology for determining quotas at Greenland, the North Atlantic Salmon Conservation Organization agreed in 1993 to base quota allocations on scientific advice from the International Council for the Exploration of the Sea," he says in an article in the *Atlantic Salmon Journal*. "To ignore that advice now makes a mockery of the whole exercise, and also reveals a fundamental flaw in the NASCO Convention. It decrees that the decisions of NASCO commissions must be taken by unanimous votes. Denmark, on behalf of Greenland, planned to vote for the proposed 77-tonne quota, but the unanimity requirement meant that an opposing vote by Canada and/or the United States would have resulted in no NASCO quota at all. Greenland would have been entirely free to set its own." Canada and the United States first argued strongly for no quota at all and then abstained from the vote.

Meanwhile, Orri Vigfusson tried to buy out the Greenland quota but failed to reach an agreement. Windsor argues that the 77-tonne quota was indeed reasonable and that countries are placing too much blame for the decline in salmon stocks on the Greenlanders.

*

For the last five years, the organization has also been trying to deal with other threats to the salmon. For example, every year in the North Atlantic, between 1,300 and 3,300 tonnes of salmon are killed but not reported to any regulatory body. These unreported catches, both legal and illegal, amount to almost half the total salmon kill in the North Atlantic. For example, New Brunswick has no mandatory reporting of catches, and the province will never have accurate scientific data or proper management of the fish harvest until all catches are carefully recorded.

Another threat to salmon is acid rain, which is killing more than 1,000 tonnes of salmon a year. Nova Scotia has permanently lost runs of about 9,000 adult salmon per year to acid rain. In Norway, Sweden and England, more than 300,000 salmon per year have been lost through the acidification of rivers. The only proper solution to this problem is the elimination of the source of the air pollution.

Furthermore, the parasite Gerodactylus salaris has killed more than 1,000 tonnes of salmon, mainly in Norwegian rivers. The parasite is believed to have been introduced from the Baltic Sea with hatchery stock which are immune to it. The parasite, which has spread into Finland, Russia and Sweden, attacks smolt, killing almost all of them. It has so far attacked 30 rivers, wiping out some genetic stocks completely.

"This is just an example of what happens as we develop fish farming and we transfer fish around," Windsor says. The organization is working now to develop geographic, political and other barriers for the transfer of farmed fish. "Gerodactylus is a wonderful example of people doing things for the best of all possible reasons — they wanted some trout in Norway from Sweden. Nobody knew about Gerodactylus. But every time we move things about there's a risk of something happening that we don't know about. And now we're into free trade, and the pressure is to have no barriers to trade." Windsor says the international community must move quickly to deal with this problem, which has the potential to destroy a river's wild salmon stocks completely.

There are other grim warnings in the Baltic, where a disease called M-74 (named after 1974, the year it was discovered) is destroying salmon stocks. Nobody knows why it has struck, but it appears to be linked to pollution in the sea. Chemicals in the sediments seeped into the water, then into the food chain and were passed on the salmon. This disease causes 90 per cent mortality in hatcheries and wild stocks. Between M-74, overfishing and a disastrous restocking program that destroyed the genetics of wild fish, the wild Baltic salmon will likely be driven into extinction in the next few years. Many scientists believe that no matter what happens, it is too late to save them. "It's an object lesson in how not to do things," Windsor declares.

Fish farming itself is seen by the international community as a major threat to world salmon stocks, with the risk of genetic pollution from farm escapees that breed in salmon rivers and the risk of disease and parasites spreading to wild stocks. In the past 10 years the production of farmed salmon in the North Atlantic has increased from 30,000 tonnes to 320,000 tonnes. In Europe, salmon farms are producing 80 times the catch of wild fish. Windsor says: "We have some rivers with 90 per cent of farmed fish in them. Is this causing genetic damage on a huge scale, or doesn't it matter? We don't know. And yet we're carrying out this huge experiment without knowing what the results will be. This could be bigger than anything we've ever heard of, or it could be something that doesn't really matter. But the fact is, if it does matter, it will be too late by the time we've found out. So we're working very hard to try to reach agreements with the salmon farming industry on reducing escapes, and on ways to treat the diseases so they don't affect the wild stocks. Eventually I think we are going to have to think about sterile fish only being used for fish farming."

The organization would also like to create "areas for the protection of wild salmon," where aquaculture is restricted or prohibited. "For 10,000 years the wild salmon has gone about its migrations, it has done its own thing, and suddenly in the last 20 years we've changed all the rules." Rivers in western Scotland saw their wild stocks decline as fish farming grew. "They can see that connection," he says. "Even if the genetic issues prove not to be serious, the disease aspects certainly are." Conservationists in Scotland and Ireland fear that wild smolt swimming by fish farms are being wiped out by sea lice. If a smolt picks up 60 or 70 lice, it will die. Windsor says the organization wants to develop, with the industry, a way of farming salmon that is not a threat to wild stocks. The salmon-farming industry is eager for reforms to be part of an international agreement that will level the playing field.

Fishing for salmon in international waters continues to be a threat. In recent years fishing boats registered in Poland and Panama have been discovered fishing salmon in the northern Norwegian Sea. During the winter of 1989, there were reports of seven vessels fishing for salmon in international waters. These ships took an estimated 630 tonnes of wild fish, more than the Greenland and Faroes fishery combined. Danish ships registered in Panama were landing wild salmon in Poland and shipping them to smokehouses in Switzerland.

The organization has been conducting diplomatic missions to try to stop this fishing. Both Panama and Poland have gotten the message because each has an incentive to stop illegal salmon fishing: Panama wants good relations with the United States and Poland wants to join the European Union. But it is difficult to monitor the illegal sea fishery. For example, in the Norwegian Sea during the winter, when it is dark 24 hours a day, illegal fishermen are usually discovered only when a ship breaks down and puts into port. In Norway and the Faroe Islands, tonnes of salmon have sometimes been found on board these illegal vessels.

*

The organization is working to build up a computer database on all aspects of salmon management. Salmon have been in our rivers for 10,000 years, but the oldest reliable data we have is from 1960. Peter Hutchinson, who is developing the salmon database, says there is a lack of historical perspective. We do know that our society has done a lot of damage to the salmon, but we are not sure how much. It's too late for us to undo the last century of damage. But in the future, until we know more about what we're doing and develop a greater historical perspective, we need to be cautious in all of our management decisions.

CHAPTER NINE

Iceland

In the blood

On my first afternoon in Iceland, I drove north from the capital, Reykjavik, to visit the country's most prolific Atlantic salmon river, the Nordura. Minutes after I left my hotel on that cool, misty afternoon in late August, I saw my first Icelandic salmon river — it was flowing beneath a four-lane highway in downtown Reykjavik.

The Ellidaar River is a small, productive salmon stream right in the heart of a busy, modern city. In its pristine pools, anglers catch an average of 1,420 salmon every year. A park with walking trails surrounds the river, and when you walk across the footbridges that span its waters, you can look down and see large salmon moving upstream or resting in pools. During the two-hour drive to the Nordura River, I crossed nine more salmon rivers. These rivers, like the Ellidaar, are all clean, full of fish and well-managed, a reflection of the way the people in this country live in their environment and manage their lives. Icelanders believe in using their natural resources — including the Atlantic salmon — to their full potential, but they respectfully, and carefully, interact with nature.

I was travelling to the Nordura with Magnus Jonasson, a former salmon fishing guide and expert angler. Jonasson was acting as a driver for his old friend James McDonough, the former president of Seagram Overseas Sales Company. McDonough is a long-time salmon fisherman and was a close friend and fishing partner of the late legendary jazzman Benny Goodman. Goodman, who lived near McDonough in Stamford, Connecticut, fished salmon rivers all over the world and had a special passion for Iceland. McDonough has been coming to Iceland for more than 20 years, and he was on his way to spend a week fishing the Nordura. "There is no fishing like this anywhere in the world," he said.

Jonasson is managing director of the company that imports Budweiser beer into Iceland (where a case of beer costs about $80). He has a Master of Arts degree in economics from Oxford University and is passionate about Iceland — especially his country's salmon rivers. Jonasson started salmon fishing as a small child. One season, he caught 445 salmon, and, in 1978, he caught 33 salmon in a single remarkable day. "It's a natural thing to fish," he says. "It's in the blood."

The Ellidaar River is a small, productive salmon stream, right in the heart of Reykjavik, Iceland. PHILIP LEE

Anglers landed 34,122 fish on Iceland's 100 salmon rivers in the 1995 season. While many countries have allowed their salmon rivers to be badly damaged or destroyed during the past hundred years, Iceland's rivers are holding their own despite poor conditions for the fish in the ocean. The salmon-fishing industry pumps an average of $32-million a year into the economy of Iceland and creates 1,000 direct jobs on the rivers and many spinoff jobs. Icelanders take their fishing — and their salmon river conservation — seriously. "We want to preserve nature, to take care of our rivers," Jonasson says.

Icelanders believe they can make money from their natural resources without destroying them. A good example of their philosophy is the eiderdown industry. Every spring, when eider ducks return to their breeding grounds on the island, eider farmers go to great lengths to protect nesting sites and make them peaceful for the birds. When the females make their nests, they pluck down from their chests to lay around their eggs. Six weeks later, when the eggs are hatched, the families of ducks leave the nests. The eider farmers move in and collect the down before the wind blows it away, and then they process it for use in clothing. This is the highest quality and most expensive down in the world. About 100 farmers make a good part of their living from these ducks, which were declared a protected species in the mid-1800s.

❋

Iceland is an island of 103,000 square kilometres that lies just south of the Arctic Circle. It is the most sparsely populated country in Europe. Two-thirds of the island's 265,000 people live in and around Reykjavik, the capital, with the rest living in small fishing towns and farming communities scattered along the island's coast. Most farmers raise sheep, cattle and Icelandic horses. Geologically, it is the youngest country in Europe, created by seabed volcanic activity 25 million years ago. This volcanic activity is still going on today, and wherever you go in Iceland you can see steam rising from natural hot springs. Iceland was the last European country to be settled. In 874, Viking adventurer Ingolfur Amarson landed on the southwest coast, at the place where Reykjavik now stands. Icelanders still speak the ancient language of the Vikings.

The Icelandic word for salmon is *lax*, a word that finds its way into the names of many rivers — Laxa i Kjos (the river that holds the record for fish caught in one season — 3,883); Laxa a Asum (the most prolific salmon river per rod in the world, this year producing 1,562 salmon in a 90-day season on just two rods); and Laxa i Adaldal (the large, rugged north coast river, the most famous of all Icelandic salmon rivers).

Reykjavik means "bay of smoke." The city was named for the steam from the many hot springs in the centre of the community, which produce heat and hot water for all homes, businesses and public buildings. This is a clean energy source; there are no smokestacks in Reykjavik. Iceland has also developed hydroelectric power on its mighty glacial rivers, which for the most part are not salmon rivers, and has a surplus of electricity.

The island's scenery is barren and breathtaking, a mixture of sharply peaked mountains, deep valleys with lush green pastures, lava fields, flat plains, glaciers and magnificent fjords. The country's coast is fissured with 250 glacial and freshwater rivers, which are fed by 1,200 lakes. The Gulf stream brings a warm climate to Iceland, running past its southern and western shores and deflecting warm air to the north. In July, the average temperatures range from 10° to 12° C while winters are relatively mild. In mid-winter, the island is in darkness 20 hours a day; in mid-summer, there are 20 hours of daylight.

Iceland's search for a new way to fish in the sea was prompted by its near-destruction of the herring fishery in the 1960s. Icelanders learned their lesson — unregulated fishing and unlimited access lead to the destruction of the resource — and herring fishing was banned completely for 10 years. In the 1970s, the country took control of its fishing grounds, declaring first a three-mile limit, then a 200-mile limit, and it began protecting them aggressively. The big cod spawning areas — thousands of square kilometres — are closed completely to fishing. Iceland was a world leader in imposing individual transferable quotas on sea fishing; government scientists develop quotas for each stock and these quotas are divided among individual fishermen, who buy, sell and trade them among themselves at the beginning of each season. Today, the country is one of the world's leading producers of seafood.

Salmon are not fished at sea. In 1932, the government realized that salmon were most valuable as a sport fish in rivers and banned salmon netting, an enlightened decision for its day. (Canada waited another 50 years before taking action against ocean netting of salmon.) All Icelandic salmon rivers are privately owned but fished by members of the public, who pay daily rod fees. The farmers who own the land bordering the rivers own the fishing rights and tend the salmon like any other farm crop. Each river is managed by its own fishing association, and there is a minimum of government interference.

Each salmon river is promoted in Iceland and around the world with photographs, details of the landscape and fishing, and a graph of accurate catch data over the past decade, including the largest salmon caught in those waters. Salmon fishing in Iceland is expensive, with most clubs charging several hundred dollars a day for fishing rights and accommodations. The productive Laxa a Asum costs about $2,000 a day to fish and is sold out five years in advance.

On these rivers, there is very little poaching. "Icelanders are not poachers," Jonasson says. "We have a respect for nature." The river owners employ wardens, but they aren't busy fighting against illegal fishing. "They are just checking to make sure everything is clean."

✻

The first Icelandic settlers soon established a republican form of government and set up regional assemblies. The general assembly, the world's first parliament, met at a place called Thingvellir on the River Ox in 930. At the end of June each year, thousands of people from all over Iceland gathered beside the river for a two-week meeting of the assembly. On my last afternoon in Iceland, I visited Thingvellir, which is about an hour's drive north of Reykjavik at the western end of a wide, flat valley and huge lake system. The meeting place of the first parliament is in a steep-sided ravine called the Almannagja. The River Ox crashes down over the cliff and runs along the floor of the ravine, which smells of the river's clean water.

The ravine itself is gradually sinking into the Atlantic Fault that runs from the southwest coast of Iceland to the northeast — right through Thingvellir. Europe and the bed of the eastern Atlantic Ocean are the eastern slab of the Atlantic Fault; North America and the Atlantic's western bed make up the western slab. At Thingvellir, more than anywhere in the world, the junction of these two slabs can be clearly seen. The day I was there, Thingvellir was almost deserted. An aura of magic and history filled the silent valley as I walked through the ravine and imagined this area filled with tents, booths and families during ancient meetings of the assembly.

In 1930, dignitaries from all over the world came to Iceland and Thingvillar to celebrate the 1,000th anniversary of the world's first parliament. In Reykjavik, Johannes Josefsson, Iceland's most famous salmon fisherman, had staked his fortune on a luxury hotel that he built to house the visiting world leaders.

✻

Josefsson was born in 1883 in northern Iceland. As a young man he became an expert in a form of wrestling called *glima*, represented Iceland in the Olympics in

The ravine at Thingvellir is gradually sinking into the Atlantic Fault that runs from the southwest coast of Iceland to the northeast. PHILIP LEE

1908 and 1912, and later joined the Ringling Brothers Circus. In 1927, Josefsson returned to Iceland a rich man.

He decided to build a hotel in the middle of Reykjavik to house the kings, queens, princes and dignitaries he knew were coming to Iceland to celebrate the 1,000th anniversary of Thingvellir. His Hotel Borg, built right in the heart of the capital, was Iceland's first top-class hotel and was a great success during the Thingvellir celebrations. But the building of the hotel put him deep in debt. He later said the fight for survival during his first 10 years at the hotel was harder than any wrestling match he had ever fought. But he and his hotel survived, and in 1940, the British army arrived in Iceland and rented his entire hotel for the duration of the Second World War. His financial troubles were over.

In 1941, he built a fishing lodge beside a deep salmon pool on the Hitara River. For the next 20 years he fished salmon on the Hitara and the Nordura, sending the fish back to the Hotel Borg. He caught about 500 salmon a year on a rod and line, landing about 10,000 fish during his years on the rivers. He died, a national hero, in 1968. His white and green lodge on the Hitara, tucked behind a rocky hill below a wild waterfall, is still used by salmon fishermen today and is leased by the Angling Club of Reykjavik. In the lobby of the Hotel Borg, the new owners have created a replica of a room at the salmon lodge, and one of Josefsson's fishing rods is displayed in a glass case on the wall.

*

An angler fishes the Hitara River at the famous lodge that was once owned by Johannes Josefsson, Iceland's most famous salmon fisherman.
PHILIP LEE

When I arrived with Jonasson and McDonough at the Nordura River lodge, south of Josefsson's famous Hitara lodge, fishermen were preparing for an afternoon and evening session on the river. The lodge was bustling; doors were opening and closing and a strong, fresh, cold wind was blowing up off the river into the lodge. A spread of food and coffee was out on the tables. Icelandic lodges are immaculately clean and comfortable, but not fancy — fishermen were packing their own lunches for the river. Outside, they were strapping fishing rods to special racks on their cars.

The Nordura lodge offers a spectacular view up a deep gorge towards a waterfall. Small springs tumble into the gorge and the hillside is a mix of deep brown volcanic rock and lush grasses. Inside the lodge, pictures of salmon flies hang on the wall. Some of them are New Brunswick flies — Nepisiguit, Miramichi Black Dose, Jock Scott. The Atlantic salmon world is smaller than you might think.

We dropped off McDonough and his river guide at the stretch of river they would be fishing that evening, which was a short drive from the lodge, just above the big waterfall. McDonough, dressed in a thick Icelandic woollen sweater and chest waders, said he was feeling 20 years younger already. Jonasson and I drove downstream just below the waterfall. There, we stood in silence in the strong wind on a cliff above the Nordura and watched wild silver salmon leap up the falls.

The rivers

Iceland is an example to the rest of the world
of how to manage Atlantic salmon fisheries.
— Dr. Wilfred Carter

Iceland has the best-managed Atlantic salmon rivers in the world. The proof is in the water — this country's clean, exquisite rivers. The system is a careful combination of restrictions and freedoms that has guided Icelandic salmon river management for more than 60 years. The basic principle is private ownership: the farmers who own the land that borders salmon waters own the fishing rights, and these fishing rights cannot be sold away from the land. Each river is managed individually with a view to conserving its salmon runs, and fishing pressure on salmon rivers is strictly controlled. The government leaves the day-to-day management of the salmon fishery to the industry.

These management principles were established in 1932 when Iceland's government banned the netting of Atlantic salmon. This crucial reform was pushed through by the farmers who owned the salmon-fishing rights on the rivers (and who also controlled Icelandic politics in the 1930s). Along with the ban on salmon netting, the Icelandic government introduced a new salmon-management policy that is a model of how river-by-river management can work. Responsibility for salmon conservation in Iceland begins with the farmers and then spreads to three groups: river associations, angling clubs and outfitters, and scientists.

Icelandic salmon legislation states that each river must be managed by a fishing association made up of the owners of the water. Each farmer who owns salmon waters is a member of the association and gets one vote. Each association must develop a management plan for its river. The association decides how many rods it wants to allow on the river, and then government scientists give final approval to this number to make sure the rivers are not being overfished. The salmon-fishing season in Iceland is limited to 90 days, but each association can decide when its river's fishing season opens and closes.

"We have adopted very strong conservation measures, the principle one being that we limit the number of rods per river," says Orri Vigfusson, the chairman of

Iceland's salmon rivers offer clean water, marvellous scenery and plenty of fish.
PHILIP LEE

the North Atlantic Salmon Fund and Iceland's leading salmon conservationist. "We limit the number of rods so that, on average, each rod catches one salmon per day. The fishing associations decide how many rods per river, but they have to apply to the national freshwater committee to approve it.

"Private ownership — that is the key. The farmers must have a fishing association on each river system, and their income is based on the number of fish caught on their pools. So all the farmers are making sure that there's no illegal fishing, no poaching, and every catch is properly recorded, because at the end of the year, that determines his share of the income." Each salmon caught in an Icelandic river is worth an average of $260 to the farmer who owns the pool.

Guarding salmon waters is the responsibility of the river associations. "The industry takes care of itself," Vigfusson says. "Everybody is taking care of their own river." River associations decide on fishing rules: there are no government regulations about fishing tackle — flies, spinners and worms are all allowed and used regularly. More than half of Iceland's salmon catch each year is hooked on worms. Some rivers set daily catch limits, others have no limits, and catch-and-release fishing is rare. Fishing on all rivers is limited to 12 hours a day.

Members of the river associations must pay a percentage of their income into a fund that helps pay for enhancement projects on the rivers. All hydroelectric projects must pay a certain amount of their income into the river-enhancement funds because it is recognized that every hydro project does some harm to the environment. Many rivers have their own hatcheries. On some pools, large plastic tubes or wooden boxes are left on the bank. When fishermen catch a big female they must put it in the tube alive and leave it in the water until it can be transported to the hatchery and stripped of its eggs. When a river needs a fishway or fish ladder to help salmon climb up over a waterfall or some other obstruction, the river association builds one.

*

River Associations lease their waters to fishing clubs, to individual fishermen or to outfitters. Vigfusson is chairman of a fishing club that has leased fishing rights on the Laxa i Adaldal in northern Iceland for 55 years. His club sells fishing days to its 120 club members and charges 15 per cent more than the lease cost to pay for overhead and enhancement projects. Then it splits river-enhancement costs with the farmers.

Fish registries and detailed catch records are mandatory on all rivers; each killed salmon is recorded in a logbook provided by the government. The angler records the weight of the fish (government-issued scales are found at all the fish registry stations), the sex, the time of the kill, the place and the type of bait used.

Scale samples are sometimes taken for study, and tags or any other marks on the fish are recorded. There are highly accurate catch records from Iceland's rivers beginning in 1940.

*

Vigfusson's club has catch records that go back 55 years. These records are now computerized and updated every week during the season. By the end of each September, the Icelandic government knows with great accuracy exactly how many salmon were caught on every river in Iceland, and where, when and how. The results are published in the newspapers and discussed by anglers all winter long.

The Angling Club of Reykjavik has 2,100 members and is the largest fishing club in Iceland. The club, which is open to all Icelanders, was formed in 1949 around the Ellidaar River that runs through the city. The club now holds leases on 10 rivers. In the fall, a catalogue detailing fishing opportunities is sent out to the members, and before Christmas they apply for licences. When the members are all booked, the club sells the rest of the fishing days. Club president Fridrik Stefansson, a lawyer with Shell Oil, says the club's primary interest is always the conservation of the rivers. "The management is built around protecting the salmon."

Club members pay fees related to the expected salmon catch — the amount goes up and down depending on whether their fishing day is in prime time or not. On the Nordura — the top river in Iceland in recent years — prices range from $200 a day to $1,100 a day per rod, depending on the date. The club recommends that two people share a rod. In prime time on the Nordura, a dozen rods are likely to land 160 to 180 salmon a week. Most of the club members are working-class Icelanders who often don't book the most expensive fishing times, but they save up for one good fishing trip a year. Fishing days on some of the smaller rivers cost $200, including the use of a lodge. If fishermen share a rod, which many do, this amounts to $100 a day. In Iceland, you can almost spend that much money buying a light lunch.

The club publishes a monthly newsletter and a quality fishing magazine three times a year. It owns its own fish hatchery and is involved in stocking programs on several rivers. In 1989, when Stefansson was vice-president of the club, he helped it permanently buy out a salmon-netting operation — one of a handful of traditional netting stations that survived the 1932 ban — in a huge glacial river into which the Nordura and several other salmon rivers flow. "Since then, that glacial river area has been in a league of its own," he says. Stefansson's club wants to try to stabilize the salmon's cycles. "We want to gently step in and stop the big drops." If the club sees a lack of smolt returning to sea, it wants to be able to fill the gap by restocking fish.

*

Iceland's Minister of Agriculture is in charge of salmon fisheries, and the Minister of Fisheries directs ocean fishing. Arni Isaksson, Iceland's Director of Freshwater Fisheries, heads the Institute of Fisheries and Aquaculture Research, located in a modern building in a Reykjavik industrial park. Isaksson studied fisheries science in Seattle before joining the institute in 1970.

Isaksson says the crucial move to save the Iceland salmon stocks was the ban on nets. There were 12 places on the island that maintained traditional salmon-netting rights, but they have now been reduced to four. The ban on nets allowed the salmon to be managed in their home rivers. He says the island's rivers thrive because they are not polluted, and the wet, cool weather provides perfect conditions for spawning salmon.

The institute provides a number of services for the farmers. Government scientists conduct research on the health of salmon rivers for the farmers who like to be kept informed about the state of their waters. Isaksson and a committee of biologists approve the number of rods that can fish each river and all catch records from salmon rivers come to the institute to be compiled. The institute directs tagging operations, which are paid for by the farmers. Isaksson works with the power companies before hydro projects begin, ensuring that the developments do not interfere with fish runs.

Setting the number of rods on a river is crucial, Isaksson says. Rod fishing kills a lot of fish, especially in Iceland, with its worm fishing and almost complete lack of catch-and-release. In a small river, a 50-per-cent harvest is not uncommon, he says. Rods will kill 50 per cent of the run in a normal year; in years of great abundance, 35 per cent; in years of low abundance, 70 to 75 per cent.

The Nordura has been the most productive river in the country in the last few years and is leased to the Angling Club of Reykjavik for $650,000 a year. In 1995, 1,670 salmon were caught on the Nordura. Just a dozen rods are allowed on the Nordura at one time, stretched over 30 to 40 kilometres of river. Such low fishing pressure is unheard of in North America. "It's a good income for the farmers," Isaksson says. "If you own a big share in the river, you don't have to do anything else."

The government is also responsible for dealing with disease prevention in the rivers; Iceland has been successful in combating diseases and takes this issue very seriously. All fishing gear must be disinfected before it is brought into the country. Anglers must present a certificate at the airport when they arrive in the country, certifying that all their equipment has been disinfected by being immersed in formaldehyde by a veterinarian. If it has not been disinfected, officials at the airport will do it for you and charge you for the service.

In the summer of 1995, the salmon disease furunculosis turned up in the Ellidaar River in Reykjavik. The government immediately declared that all fishing gear used in the Ellidaar had to be disinfected. A more drastic decision was made a few weeks later. The government decided to destroy most of the fish in the Ellidaar and start from scratch to try to eliminate the disease.

The institute receives half of its funding from private sources. "The government is not very willing to spend public money on salmon matters," Isaksson points out. The farmers make money on salmon and the government expects them to put something back into their resource. The government pays about $760,000 for the institute, and the same amount is received from salmon water owners and power companies. "Everything here is very sweet and small," Isaksson says. There is very little bureaucracy in the department. Isaksson deals directly with the minister and he also works with the biologists in the field. The institute has about 20 employees.

Over the past 50 years, the number of salmon in Iceland's rivers has fluctuated. The North Atlantic in recent years has not been able to produce enough adult salmon, Isaksson says. He has two theories about what may be contributing to the problem: The water has been too cold for salmon, and there has been a "phenomenal explosion" in the number of seals and whales feeding on fish — both salmon and the salmon's food supply. Other marine species have also been having problems — the cod and capelin, for example. Their problems could be linked with the salmon's problems, and scientists need to look for patterns. Isaksson argues that there must be more co-operation among marine scientists working in different fields to try to find answers for the problems in the sea. "We know the freshwater phase of salmon really well. We don't know as much about the marine environment. We are trying to tie all the links together."

Orri Vigfusson

On December 31, 1999, Orri Vigfusson plans to throw a party to celebrate the completion of his mission. A dark-haired Viking with a flair for the dramatic, Vigfusson has rented the famous Bolshoi Theatre in Moscow for the event. There, he plans to announce the salvation of the world's Atlantic salmon stocks. "We are halfway there," says Vigfusson, 53. "I have been working for five years. I will work another five years and then I will give it up. We are going to deliver the world's salmon stocks healthy into the next millenium. That's our mission."

The headquarters for Vigfusson's mission is a small, plain office above a strip mall in downtown Reykjavik. The office is also Vigfusson's place of business: he exports Icy Vodka, a high-quality vodka his company distills with pure Icelandic water. But for the past five years, Vigfusson's office has turned into a salmon conservation command post — every piece of paper, piled high on his cluttered desk, has to do with salmon. For five years, Vigfusson has been working more than full-time on his salmon mission. "He is obsessed," says his assistant Vilhelm Steinsen. "He spends more time abroad working on his campaign than he does in Iceland." Fridrik Stefansson puts it simply: "Orri is on a one-man mission to save the salmon of the world."

Vigfusson founded the North Atlantic Salmon Fund in the fall of 1989 with the intention of stopping the high-seas Atlantic salmon fishery off West Greenland and the Faroe Islands, which has been killing salmon from countries all over the North Atlantic before they return to their home rivers to spawn. Vigfusson says his campaign to stop the high-seas salmon fishery is the single most important salmon conservation battle.

Orri Vigfusson on the Laxa i Adaldal, Iceland: leading the charge against the high seas slaughter of salmon.
Courtesy of Orri Vigfusson

For years, conservationists worked to control these fisheries through international negotiations, but these negotiations were too slow and ineffective for Vigfusson. He had a new approach based on the philosophy according to which the ocean fish stocks are managed in Iceland: quotas, set by scientists, for each species of fish, divided among fishermen who are allowed to buy, sell and trade them among themselves.

Vigfusson considered the Greenland and Faroe Islands salmon fisheries to be the property of those fishermen — and he planned to buy them out. "We developed this concept which we think is fair," he says. "We simply paid them not to fish." The money was important, an approach based on the assumption that Greenlanders and Faroese have the right to fish. Atlantic salmon feed and grow in their waters and so they have a right to be paid for their contribution to the resource.

*

In the fall of 1995 in Dublin, Ireland, Vigfusson addressed a salmon dinner of 550 people who paid $120 a plate to help raise money for his fund. With an auction at the end of the dinner, another $100,000 was raised for the high-seas buyout. Vigfusson told the gathering that the Atlantic salmon is an important part of Icelandic culture — only the country's language, sagas and poetry are more important. He continued:

> The North Atlantic Salmon Fund was set up to develop a new concept and principle in fisheries. We felt that the good work done by managers and scientists in fresh water was worth nothing if fish did not feed safely in the sea and then migrate back to their rivers of origin. No matter what magnificent management schemes, enhancement programs and conservation policies we have in our home rivers, these will yield very little if we do not pay full attention to the problems at sea. There is no doubt that freeing the seas of both legal and illegal fishing and irresponsible industrial fishing, which attacks the very base of the food chain, would pay vast dividends in the future. Our long-term goal is to develop new fisheries and alternative employment for fishermen who voluntarily give up their right to fish for salmon.

The North Atlantic Salmon Fund purchased the Faroe Islands' 550-tonne salmon quota in 1991. In an agreement reached with the help of a mediator, it paid the fishermen 27 per cent of the estimated landed value of the quota — about $800,000. The agreement has been renewed every year since then. "We hold unlimited purchase options so long as we can foot the bill," Vigfusson says. He bought out the Greenland fishery in 1993 and 1994 at a cost of about $400,000 a year. Negotiations failed in 1995, and the Greenland fishing fleet, which at its peak involved 1,200 small boats, caught about 70 tonnes of fish. The fund has options to buy future Greenland quotas, which "we have every intention of exercising," he says.

In addition to buying quotas, Vigfusson is seeking alternative sources of income for these fishermen, encouraging the harvest of lumpfish, sea urchins and crab. "We have to make sure that these guys who give up the fishery are better off, that they have a much better life," Vigfusson says. "We have to find alternative employment for them. Otherwise, we will never be able to do this successfully in the long term. They will never give up their salmon quota if they have a lower standard of living. And it's wrong to expect them to."

Golf legend Jack Nicklaus (left) and Orri Vigfusson on the Laxa i Adaldal, Iceland.

Courtesy of Orri Vigfusson

In the 1980s, before Vigfusson's fund began working, 80 per cent of the 3.4 million salmon being harvested every year in the North Atlantic were taken by the high-seas fishery. Although that number has now been reduced to 55 per cent, it is still too high. "In the last 20 years, the most damaging factor has been the overharvesting at sea. All fishing of mixed stocks of salmon is bad. It's bad from a biological point of view because you can't control the fishery from any one particular river system. Salmon should be harvested at the rivers where they return, where you can control stocks."

The fund has raised $4-million ($1.2-million in North America and $2.8-million in Europe) and spent $4.5-million. Vigfusson borrowed the extra $500,000 himself. He says he needs to raise another $8-million over the next five years. "We have not been able to raise enough money to meet our obligations," he says. "Not enough people have been helping us. Some people are saying, 'Let the rich pay for it on their own.' But with more salmon, everybody benefits, not just the rich." Will stopping the high-seas fishery alone save the salmon from extinction? "No, but I think that's the single most important factor."

*

Vigfusson travels constantly: to Washington to seek support from the government, to Dublin and Copenhagen for fund-raising dinners, to France to support the removal of dams from the Loire and Allier salmon rivers, to fund-raising events in Spain. Vigfusson's organization is now also promoting salmon conservation around the world, urging countries to stop fishing salmon at sea in their home waters. Vigfusson is especially concerned about salmon drift-netting operations off the English and Irish coasts. He is also encouraging countries to develop the Icelandic system of individual river management.

The president of Iceland has awarded Vigfusson the Order of the Falcon — an Icelandic knighthood — for his work in salmon conservation. And in April, 1995, at St. James's Palace in London, the royal palace next door to Buckingham Palace, Prince Charles presented Vigfusson with a life-sized sculpture of salmon on the spawning grounds. At the reception, attended by 100 conservationists from around the world, Prince Charles thanked Vigfusson for his tireless campaign to protect the Atlantic salmon.

❋

St. James's Palace is a long journey from the small fishing community on the north coast of Iceland where Vigfusson grew up. His father was involved in the herring-processing industry, shipping the fish to Scandinavian countries and the Soviet Union, and he also owned shares in herring-fishing boats. When the herring stocks crashed from overfishing in the 1960s, Vigfusson's family suffered, and he learned a lesson in conservation he never forgot.

Prince Charles presents Orri Vigfusson with this sculpture of spawning salmon to honour his conservation campaign. Courtesy of Orri Vigfusson

Vigfusson had studied economics and marketing at the City of London College in London, England. There he met Unnur Kristinsdottir, his future wife, who was also from a small town on the north coast of Iceland. She introduced Vigfusson to salmon fishing at her family's lodge on the famous Laxa i Adaldal River on the island's north coast. In the late 1960s, Vigfusson joined the river's fishing club. "I very quickly became involved with the farmers in enhancement projects," he says. For the past 10 years, he has been the chairman of the 120-member fishing club on the Laxa i Adaldal.

He starting working full-time for the salmon in September, 1989. The idea of trying to stop the high-seas fishery came to him when he discovered that the salmon smolt he was tagging and releasing in his river were being killed in nets off Greenland. He asked himself: Why didn't our club members get the salmon? Why were they being caught on the high seas?

The fund began with members of his club, who were celebrating their 50th anniversary on the river. They decided to make the North Atlantic Salmon Fund their gift to the river. Then Vigfusson started contacting his friends in England, Sweden and North America. In November, 1989, he came to the Atlantic Salmon Federation's annual dinner at the Plaza Hotel in New York City and presented his plans. He was made a director of the federation and the campaign was launched. The federation has contributed $650,000 (U.S.) to the cause. The United States State Department has put in $300,000, the National Fish and Wildlife Federation, $250,000. Dr. Wilfred Carter has worked closely with Vigfusson during negotiations with fishermen in Greenland and the Faroes.

*

Vigfusson's wife and two children all support his salmon campaign. His wife and daughter were visiting the Big Laxa when I met Vigfusson in Reykjavik in the summer of 1995. His son, Vigfus, 23, is an expert angler and fishing guide who is working his way through university with money he makes on the rivers.

On a cool evening in August, 1995, Vigfusson and I travelled to the Sogid River on the south coast of Iceland to visit Vigfus — and do some business. Vigfus was guiding Nigel Hawkins, the president of the Icelandic Angling Club, which was established in London, England, in 1862. When we arrived at the lodge, Hawkins and Vigfus were just climbing up the bank after an evening of fishing on the deep, wide river, the largest clear-water river in Iceland. At the lodge, as the sun set, Vigfusson and Hawkins talked about the fund's conservation campaign and future fund-raising events. All membership fees from Hawkins' club are paid directly into the North Atlantic Salmon Fund. In his letter to potential club members, Hawkins writes: "Orri Vigfusson has dedicated himself to the cause of the Atlantic salmon and has singlehandedly done more for their future than any other man. This policy of buying out quotas is already providing results and over the next two or three years is expected to have a huge impact on reversing the horrendous decline of our wild stocks of salmon." We drove back to Reykjavik in the dark — a typical evening for Vigfusson during fishing season in Iceland.

*

Vigfusson is a force to be reckoned with — and so are the Greenlanders. Einar Lemche, the representative of the Greenland Home Rule Government, laid it on the line when he spoke to the North Atlantic Salmon Conservation Organization's annual meeting in Oslo, Norway, in June, 1994: "Though we may from time to time refrain from utilizing our family silver, we are certainly not going to sell it."

Small-boat fishermen in Greenland do not believe the scientists who tell them the number of Atlantic salmon in the sea has reached an all-time low. Their skepticism about science is one of the main reasons the negotiations to pay them not to fish failed in 1995. Vigfusson says the Greenlanders believe strongly that the scientific formula that gives them their salmon quotas is producing numbers that are far too low. "Some of them are saying, 'We must fish, we must show the scientists there's lots more salmon in the ocean.'" In all this talk about quotas and buyouts, it's easy to forget that the Greenlanders have their own way of life to defend. Their traditional hunting and fishing lifestyle may be as threatened as the wild Atlantic salmon.

Vigfusson's conservation campaign has its critics, among them Dr. Malcolm Windsor of the North Atlantic Salmon Conservation Organization. He says the buyouts are no more than a short-term measure in a fishery that needs long-term rational management. "It's not a buyout," he says. "It's a rental or a lease, because the Greenlanders have said that they might refrain from using the family silver from time to time but they will never sell it. So it means you will have to negotiate the leases in perpetuity."

*

At a meeting of the International Council for the Exploration of the Sea two years ago in Edinburgh, Scotland, Einar Lemche presented a powerful argument on behalf of the Greenland fishermen, who, he said, feel cornered. Greenland is a remote country of 40,000 people, a mixture of Danes and native Greenlanders, under Danish rule. Communication on the international stage is difficult for them, and there is a great cultural divide.

"This is not only a matter of biology, not only a matter of getting as much as possible in a negotiation," Lemche said. "We feel that for our opponents, this is also an attempt to undermine our lifestyle, our means of subsistence. Wealthy city dwellers far away force us to fight an endless battle to be able to continue our traditional whale hunt. Wealthy city dwellers far away prevent us from marketing one of our few plentiful resources, seal skins. Wealthy city dwellers far away are now trying to impose a complete end to commercial salmon fishing in our country. In short: If you want to promote dialogue on salmon, you need to convince us that Greenland is not the target of yet another crusade based on cultural imperialism."

Lemche pointed out that salmon in Greenland are harvested by small-boat fishermen. Cod have virtually disappeared, and salmon are still an important cash harvest. "Some are even requesting the complete cessation of fishing of every salmon which is not eaten by 'native' Greenlanders. In other words, they are requesting our poor fisherman to refrain from utilizing the fish under his boat in order to increase the number that rich anglers can take in other countries."

Paviaaraq Heilman of the Greenland Hunter's and Fisherman's Organization told the Edinburgh group that a dramatic increase in the whale and seal population is killing salmon much faster than their nets. And the Greenlanders have another point to make: In the past 10 years their salmon quota has dropped from 2,500 tonnes to 77 tonnes. In 1993 and 1994 there was no salmon fishing in these waters at all. Why are the world's salmon stocks still in free fall? "In the past, even substantial cutbacks in the Greenland catches have not improved the situation in

Greenland salmon fishermen work in small boats like these.
ATLANTIC SALMON FEDERATION

home waters," Lemche said. "There is no reason to anticipate that future cutbacks would."

Anglers and conservationists on both sides of the Atlantic have heaped praise and dollars on Vigfusson's North Atlantic Salmon Fund. But Dr. Malcolm Windsor says the buyouts are making it more difficult to negotiate a reduction in salmon quotas on the high seas. Greenlanders accepted a mechanism to set scientific quotas for salmon before the buyout began. In the Faroes, there wasn't a mechanism in place because the science wasn't advanced enough. Now that the Faroese are getting paid for not fishing, they aren't interested in setting a lower quota.

The buyouts, Windsor says, "are not a conservation measure. It's a transfer of catch from one group of fishermen to another. Orri Vigfusson would pay the Greenlanders not to take the fish because he wants them back for anglers. And there's nothing wrong with that, but it's often presented as a conservation measure, which it isn't. It's simply a transfer of the catch from one sector to another."

Windsor says that while he is not opposed to Vigfusson's buyout campaign, "it probably deflects funds and attention from the real problems. The real problems with the salmon in the North Atlantic are not the interceptory fisheries. The problem lies at home, really. We would prefer to see resources put into where the problems are. There's an inherent problem here in perception, that we can solve conservation problems by methods like this. We can't."

For example, rivers in the Western Highlands of Scotland are suffering from a major collapse of salmon and sea trout; the stocks are extinct in some areas. The sea trout are coastal fish; the salmon are grilse that return to the rivers after one winter at sea. So they're out feeding in the Atlantic, but not as far away as Greenland. Fisheries associations in these areas have been contributing to the buyout of the Greenland fisheries. Dr. Peter Hutchinson says the people there are beginning to ask themselves if this is the right way to spend their money. "Our problems lie in home waters. We've got fish farming. We've got things going on in fresh water. What we really should be doing is putting our money into a trust here and getting some work done on the rivers locally. I think that in Scotland, certainly in that area, there are people now who are beginning to question the buyout and are beginning to say that the biggest threat to these stocks is not, and has probably never been, the Greenland fishery, because our stocks don't go there. Our problems lie at home on our own doorsteps."

✻

Vigfusson says Windsor's North Atlantic Salmon Conservation Organization is an important conservation force and he is proud it was created in Reykjavik. He agrees with Windsor's assertion that the buyouts will have to be long-term agreements. That's why his fund is working to try to find alternative fisheries for the salmon netsmen.

"We want to have more salmon and we want these netsmen to have a better living. Everybody must benefit. We are not in this game just for one year. This is what belongs to them. It belongs to their country. So they should own it in the future. We simply have to pay them a grazing fee." These are salmon that are feeding and growing in Greenland and Faroe Islands waters. "Why shouldn't they get paid for that?" He hopes that when he retires from the fund in five years governments will step in and pay this "grazing fee."

Vigfusson says it was not until he began to spend money that members of the North Atlantic Salmon Conservation Organization realized they should be doing more work. "I think it's good to put a monetary value on the resource. When you do that you have much more sensible solutions. The quotas in Greenland were not reduced until we had money on the table."

The Greenland salmon fishery once killed 750,000 Atlantic salmon a year.
ATLANTIC SALMON FEDERATION

As to the charge that he is buying quotas so that the fish can be caught on rods and lines in rivers, Vigfusson's answer is simple: "Yes, we are." But he points out that anglers are beginning to catch and release fish. And a salmon is worth much more on a rod and line than in a net. Why does it make sense to net salmon when you can get $1,000 for that same fish in a river? And he points out that rivers will never be well managed if an undetermined number of fish are killed in the ocean before they return to home waters.

Vigfusson is lobbying hard to convince the British and Irish governments to stop allowing unregulated drift-netting for salmon off their coasts. In recent years the drift-netters have been catching more fish because Vigfusson's buyout at Greenland and the Faroes has allowed more fish to return to the coastline. Why should Greenlanders and the Faroese have to accept a total quota, Vigfusson asks, when there's no quota in the home waters of countries like England and Ireland? On the market in London, the best price for a netted salmon is $30 a fish. "Instead, here in Iceland, we're getting $1,000 a fish in the rivers. It's ludicrous," Vigfusson says.

*

On his 50th birthday, Vigfusson found out he had intestinal cancer. Barely missing a beat in his conservation campaign, he had surgery and spent a year in a chemotherapy program. "I wasn't frightened," he says. "I had always been afraid of hospitals. But at the national hospital, it was my fishing partners who operated — I had four fishing partners in the operating room with me." When he turned 52 in July of 1994, Vigfusson was given a clean bill of health, exactly 2,000 days before his deadline — the year 2000. He doesn't seem to get tired. "It's my passion. I like to work and I like to get things done."

CHAPTER TEN

The Rescuers

Heroes

Men, women and children came to New Brunswick's Big Salmon River late in the morning as the October sun strained through the fog. About 40 people — most of them wearing waders or rubber boots, some of them playing hooky from work — gathered by the footbridge that crosses the river just downstream from the Amateur Pool. They were waiting for their salmon — their "babies" — to arrive. The Amateur Pool is the river's last deep salmon pool before its clear waters run down over a rocky plain, cross a beach and empty into the Bay of Fundy east of the fishing village of St. Martins.

At 1 p.m. the flatbed truck arrived, carrying a load of 80 large, live Atlantic salmon in seven big, square grey tanks filled with a mixture of ocean and fresh water. Beulah Armstrong, the president of the Big Salmon River Angling Association and the leader of this small army of volunteers on the riverbank, arrived with the truck and pulled on a pair of brown chest waders. She looked worried. "We've got some sick babies," she said.

The fish, which Armstrong's association had raised in a sea cage to help restock the Big Salmon River, were covered with sea lice. Like other farmed fish in the Bay of Fundy in the fall of 1995, the Big Salmon River fish were hit with a severe case of lice. But unlike the farmed fish that have to remain at sea, these salmon were about to get some relief — sea lice drop off and die in fresh water. Armstrong walked by the truck and scooped up her granddaughter, Olivia. "She had to come and see Grammy's fish," she said.

A hose and pump filled two empty tanks with river water. One by one the fish in the saltwater

The scene from the footbridge above the Amateur Pool at Big Salmon River the day the "babies" were put in the water. DAVID NICKERSON

Beulah Armstrong searches for salmon in the waters of the Amateur Pool on the Big Salmon River, New Brunswick. DAVID NICKERSON

tanks on the trucks were lifted by net into the fresh-water tanks. Meanwhile, the volunteers formed a line, a human chain, along a narrow dirt path that leads through the woods down to the Amateur Pool. The people who were wearing waders, with Armstrong in the lead, stood in a line in the river at the foot of the pool.

The salmon were taken from the fresh water and put, one at a time, head first, into a water-filled "boot" — half of an inner tube tied off at one end with a piece of rope. The first salmon was quickly passed along the human chain down the path into the hands of Armstrong, who released the fish into the water and gently began to revive it, moving it back and forth in the current. When the salmon swam into the pool with a hard thrust of its tail, cheers rang out across the pool. And then they went back to work. "We're going to need everybody with waders to be working fish," Armstrong shouted.

For the next two hours they went at it hard, lugging the four-year-old fish, which weighed between 13 and 20 pounds, down into the river. Volunteers were scattered around the pool, some up to their waists in the water, reviving fish. One man kept his arms in the cold water for 30 minutes trying to save the life of a fish that was stressed by the trip to the river and a bad case of sea lice. By 3 p.m. all the fish were in the river, 30 of them with tags attached. Only four salmon had died.

Armstrong and her volunteers stood on a rock ledge beside the pool and watched the salmon roll and swirl as they adjusted to the river water. Then they headed for their cars. The next day they planned to put 150 more salmon in a pool farther upriver. Bonnie James and her husband Brian stayed behind in their trailer, which was parked at a campsite beside the river. They were planning to stay there for a week to guard their babies. "We'll be doing a 24-hour watch here," she said.

*

Armstrong and her volunteer army are fighting the extinction of the salmon runs in their river. The situation is critical: in 1995, only 110 wild fish returned. In 1985, 4,500 salmon returned to Big Salmon River to spawn. By 1986, anglers on the Big Salmon knew there was a problem: The number of fish returning to the river suddenly dropped by half. In 1987, the run again dropped by half. At that time, Armstrong wrote to then-federal minister of fisheries Tom Siddon, the man ultimately responsible for salmon conservation. She says that she was deeply concerned about the future of their river and suggested that the federal government look for the cause of the alarming decline in the Bay of Fundy salmon stocks.

Members of the Big Salmon River Angling Association put one of their "babies" into a water-filled boot for transport to the river. DAVID NICKERSON

Siddon responded first through his executive assistant, who wrote a patronizing letter discounting Armstrong's suggestions of possible reasons why the fish were dying at sea. The letter implied that the problem was not serious by noting: "The lobster population in the immediate area is thriving." Armstrong wasn't sure what the lobster had to do with her salmon. When Siddon himself wrote to Armstrong in January, 1988, he too refused to acknowledge the importance of her concern: "We expect salmon stocks in your river will return to more normal levels in 1988," the minister said. Research budgets were strained, he added, promising to reassess the situation in the future. "No one believed us that there was a problem," Armstrong says. "It just started dropping by half every year." In 1988, despite Siddon's groundless optimism, the run was down to 500 fish.

In May, 1989, Siddon wrote to Armstrong and told her he remained committed to salmon enhancement and thanked her for her devotion to the river. In 1989, the run rebounded to 1,200 fish, but then it dropped off sharply again. In 1990, 500 fish returned, and in 1991 and 1992, 150 fish returned each year. The river has been closed for all fishing since 1992.

*

The Big Salmon River has clear water and deep pools. It runs around large boulders scattered along its 24-kilometre valley. "It's a classically beautiful river," says Dr. Alex Bielak. "It's as if God had laid a mosaic and you can see every stone on the bottom, every salmon. The tragedy is that we haven't had that many salmon." At the mouth of the river, the water flows out of a gorge into the Bay of Fundy. On the beach, you could once see the salmon running in with the tide. "When the fish were out here ready to come in they would start jumping when the tide started coming in," Armstrong says. "I can remember us walking way out on the sand, and we could count, and at one time there'd be 20 fish in the air. You'd never dream it would come down to this."

A human chain of volunteers carries the salmon to the Big Salmon River's Amateur Pool.
DAVID NICKERSON

Roderick Haig-Brown wrote about the Big Salmon River's special breed of salmon in *Bright Waters, Bright Fish*. He said these fish "spawn for the first time as grilse, migrate and go only a short distance from the mouth of the Bay of Fundy (keeping safely south of the Newfoundland nets), and return every year from then on, up to five or even eight times, to spawn again." Haig-Brown thought if other river stocks could gain some of these genetic characteristics, then salmon could perhaps be saved from the harrowing journey of crossing the Atlantic. Now, the Big Salmon and all of the Bay of Fundy salmon rivers are in serious trouble.

The Big Salmon River valley was once owned by the Hearst Corporation, which operated a timber operation upriver and a mill at the mouth of the river. In the 1940s and 1950s, James Catt, the province's regional supervisor of fish culture, fought to restore the salmon runs, which were dying because of a dam at the mouth of the river. He helped to establish a fishway in the dam and restocked the headwaters of the river. Catt died while fly fishing there on September 16, 1959. A plaque in his memory was erected in a field across the footbridge near the Amateur Pool.

No one is sure why the Big Salmon fish are dying at sea, and the federal government — despite its constitutional responsibility for salmon conservation — has never launched a significant scientific program to try to find out what's happening. One thing is for certain: The problems lie outside Big Salmon's waters. "It's an exquisite river for young salmon. That's not the problem," says John Blenis, a provincial fish and wildlife technician who has been working with Armstrong's group.

✻

The Big Salmon River Angling Association, which Armstrong has led for 10 years, grew out of a group of people who camped at the mouth of the river every summer. The camping area runs along a high bank above the river gorge, and from there you can see the beach and the Bay of Fundy and look down into the clear waters of the river. Beulah Armstrong, her husband Eddie, a truck driver, and their three children started camping at the river in the summer of 1968, where they fished for salmon while the children swam in the river and played on the banks. Every spring, members of the association cleaned up the campsite, cut the grass, piped in spring water and built outhouses. This was a community of working-class salmon anglers — no big spenders here, just people who loved the river.

In January, 1993, members of the association met at the Armstrongs' home in Willow Grove, a suburb of Saint John. There were only 13 members of the association left — their numbers had dropped as the fish runs diminished. It was time to make a decision: Should they disband the association and give up on the river, or do something to try to save the salmon? They weren't ready to give up — but they needed a new plan of action. The federal government had been stocking young salmon in the river, but a yearly supply of smolt wasn't helping the runs. Armstrong believed that something was happening to the smolt as they migrated from the river into the Bay of Fundy.

A few years earlier, Tom Pettigrew, the province's regional biologist based in Hampton, had proposed trying to capture the Big Salmon smolt as they left the river and raising them in a sea cage for one winter. Pettigrew had prepared a proposal to do this, but governments weren't interested in pursuing his plan. That January, Armstrong suggested to the members of the Big Salmon River Angling Association that they try to resurrect the Pettigrew plan. She went to the Atlantic Salmon Federation in St. Andrews to see if it would help fund the project. The federation said it couldn't help her at that time.

A determined Armstrong changed gears and went to see Doug Aitken, the manager of the federal government's Saint John Fish Hatchery. She suggested that her association take 500 hatchery fish of Big Salmon River origin and raise them in a sea cage. The association would then put the sexually mature fish into the river to spawn. The fish would seed the river and then, they hoped, go out to sea and return to spawn again. This type of restocking had never been done before.

A large Atlantic salmon swims away after being released into the Big Salmon River.
DAVID NICKERSON

Aitken saw no problem with her plan, so Armstrong ran it past Pettigrew, who told her to ask for 2,000 fish. Armstrong telephoned Dr. John Ritter at his Halifax Department of Fisheries and Oceans office and he agreed to provide the young fish. Then she went to governments looking for the $60,000 or so they needed to buy a sea cage and raise the fish at the New Brunswick Salmon Growers Association Demonstration Farm in Lime Kiln Bay. She was turned down flat.

In April, 1993, the association held another meeting. "We said, 'We're going to go for it on our own. We're going to raise the money and do it,'"

Armstrong recalls. They needed to raise $37,000 right away. They held a lottery and sold 2,300 $10 tickets. They were on their way. On May 25, 1993, 2,000 Big Salmon River smolt were delivered to a cage at Lime Kiln Bay. Since then they have run two more lotteries and raised more than $50,000.

Meanwhile, in the sea cage, their fish grew from two and a half ounces to between five and nine pounds in the first year. In the fall of 1994, members of the association sorted the 1,248 fish that had survived in the cage. They selected 379 sexually mature fish. On October 6, 1994, Armstrong and six other volunteers placed the fish in the river. "We carried those fish from the top of the bank down into the river one by one," she says. "The last 40 were put in with the aid of truck lights. We had a long day." About 100 fish were put into a tributary called Falls Brook above a 175-foot waterfall. No wild salmon can migrate above the falls and these fish would test Armstrong's theory that the fish from the cage would spawn in the river.

Three fish died during the stocking in 1994 and about a dozen were lost to poachers the first night. The association members guarded the river 24 hours a day for three weeks and they watched the fish follow their instinct and began to swim upriver to spawn. Wild fish and caged fish were seen mingling together on the spawning grounds. "The instinct of moving up to spawn is there even though those fish have never been in the salmon river," Armstrong says.

In the summer of 1995, Armstrong and members of the association, along with provincial biologists, went above the waterfall to search for baby salmon. The first fish they found was a young salmon — the tributary was full of healthy young salmon. The stocking plan had worked. "Beulah was tickled pink," says John Blenis. Now they are waiting to see if a fish raised in the sea cage will return to the river to spawn for a second time.

*

The Big Salmon River Angling Association has more than 60 members and has put one million eggs into the river. "We have a lot of members now who don't even fish," Armstrong says. "They just have heard of the project and they want to back it. It's really encouraging. This is about trying to save a strain of Atlantic salmon." But, despite valuable moral support from individual provincial and federal biologists, the association continues to fight alone to save this river. Instead of launching a campaign to preserve it, the provincial government is planning to spend millions of dollars to build a paved road down to the river and erect a bridge across its mouth as part of the Fundy Trail project. "It's just a boondoggle," says Jack Fenety.

The Big Salmon River, with crystal-clear waters, is one of the most beautiful small salmon rivers in the world.
ATLANTIC SALMON FEDERATION

Beulah Armstrong and the volunteers who spent that October afternoon at Big Salmon River in 1995 are heroes. For two hours, Blenis, dressed in rain gear, scooped the heavy fish out of the tanks with a net — it takes a strong back to do this — and passed them down to a volunteer on the ground. John Waugh, a resident of Lepreau and a new volunteer with the association, spent two hours with his arms immersed past his elbows in the cold river water reviving fish. He is a salmon angler who fished the Big Salmon the last year it was open. "Everybody else seemed to be playing their part, so I thought I'd do mine," he says.

"Our hearts are with the river and with the fish," says Beulah Armstrong. "If we could just see them come back. I would like to be able to say I could go fishing again. I hope I live to see the day."

The resurrection of the Nepisiguit

In 1969, all the young Atlantic salmon in the Nepisiguit River died in a poisonous flood. That year, and for the next five years, adult Nepisiguit salmon returning from their feeding grounds in the North Atlantic turned away from the river mouth, blocked from their spawning grounds by toxic river water. This environmental disaster on one of northern New Brunswick's finest salmon rivers was caused by heavy, acidic rains that joined forces with an open-pit lead and zinc mine operated by Brunswick Mining. The rains combined with waste sulphide piled in a swampy area near the mine to produce a lethal runoff.

In 1976, Robert Baker, a devoted salmon fisherman and the head of industrial engineering at the mine, decided to try to bring the river back from the dead. He and a group of about 20 salmon anglers formed the Nepisiguit Salmon Association. "Other than my love for fish, it was also that I felt a responsibility as part of the mining company to do something about it," he says. Baker launched a remarkable restoration program on this river. The results are in the water.

Upper Pabineau Falls on the Nepisiguit River in northern New Brunswick, which was rescued by a tenacious band of volunteers.
ATLANTIC SALMON FEDERATION

In the fall of 1995, Baker was out after an October rainstorm counting salmon spawning nests in the river, which now has clean water and steady runs of fish. Baker and the dedicated members of the Nepisiguit Salmon Association have quietly gone about their work and have shown how powerful river communities can be when they take decisive action. "We've had enough successes to keep a person encouraged and show that it can be done." Baker is a firm believer that people must take ownership of their rivers and guard against the kind of disaster that struck the Nepisiguit in 1969. "The people have to play a larger part in protecting their own resource."

*

The Nepisiguit flows 130 kilometres from the Nepisiguit Lakes through the city of Bathurst into the Bay of Chaleur. Only the last 28 kilometres of the river are accessible to salmon. The rest of the river is blocked by an impassible 30-metre double falls. The Indian name for the river is "winkkapiguwick" — place of troubled waters — for the river drops quickly from the falls 215 metres into the sea through many salmon pools littered with large boulders. Most of the river is open for public fishing. Just below the falls, the Stone Consolidated forestry company has riparian rights and leases the fishing rights to a local group, and one outfitter has private water and runs the Nepisiguit River Camps. Most other riparian waters are held by Stone Consolidated, but the company lets the public fish for salmon.

*

Robert Baker, a Nova Scotia native who grew up fishing the Annapolis River with his father, came to New Brunswick after working at a mining operation in Tilt Cove on the northeast coast of Newfoundland. He started fishing the Nepisiguit four years before the disaster and so got a taste of the river's potential.

Baker says Brunswick Mining took immediate action to try to clean up the river water as soon as the company realized what had happened. "At that time there were no regulations about the piling of waste sulphide and no knowledge of what would happen. After that happened, the company spent a considerable amount of money in order to clean up the waste piles and move them out of the way and lime the water going out of there. They moved quickly once they found out what happened. It's unfortunate that there weren't regulations already in place."

It took five years for the water quality to improve and for the river system to flush itself out. "Then, of course, there were hardly any fish to come back, except a few in the tributaries," Baker says. In 1974, a few salmon, descendants of the survivors in the tributaries, came back to the river. Two years later, Baker and his team of volunteers took action.

During these early years, the association worked with the Department of Fisheries and Oceans, stocking the river with salmon smolt, two-year-old fish that were ready to leave the river and go to sea. Because there were few Nepisiguit salmon left, the association obtained salmon eggs from Restigouche and Miramichi stock. As soon as these fish started returning to the river, all brood stock was taken from the river to try to keep the new genetic Nepisiguit strain as pure as possible. The Restigouche and Miramichi fish Baker's group chose were so-called "early run" salmon that come upriver in May. The Nepisiguit had never had any of these prized early-run fish, even before the disaster of 1969. It does now.

Soon the association moved away from stocking smolt. Baker says it didn't make sense to put tame fish into the river just before they were supposed to go out to sea. They decided to start stocking salmon fry, newly hatched fish, to give them a chance to grow a couple of years in the river and become wild before going out to sea.

In 1981, the association decided it had to start its own salmon-enhancement program if it was going to save the river. The Department of Fisheries and Oceans was moving too slowly for Mr. Baker's group. The association built a 215-metre-long counting fence on the river and started monitoring fish runs and collecting brood stock for restocking programs. At the same time, Baker's group began doing juvenile fish surveys on the river. These projects, financed by the Department of Fisheries and Oceans, federal job-creation programs and private donations, cost $1.5-million. The program worked. In 1981, 312 salmon went past the counting fence, and in 1988, 4,268 salmon were counted there.

In 1985, the association joined forces with Brunswick Mining's Noranda Research division to put "streamside incubation boxes" in Pabineau Brook, a tributary of the Nepisiguit. These boxes allow salmon eggs to hatch right in the river's water while protecting the fish from predators. This was the first time this technology had been tested on Atlantic salmon. "You put the eggs in the boxes in early spring and allow the water from the river to flow over and through them until they hatch out and become swim-up fry. The moment they become swim-up fry you put them out in the stream. They are as close to a natural fish as you could possibly get." The first year, they hatched 25,000 eggs, and 96 per cent survived, whereas in nature, about 20 per cent of the eggs survive. The next year they hatched a box of 50,000 eggs. In 1987, they moved to the base of Consolidated Bathurst's hydro dam, which is located at the large falls, and set up seven boxes that hatched 350,000 eggs. The association has hatched more than two million young salmon in these boxes. All told, since 1981, the association has put more than seven million young salmon into the Nepisiguit.

Like all salmon rivers, the Nepisiguit has been hurt by poor survival rates for salmon in the ocean. "In spite of the low survivals, in spite of the high-seas fishery off of Greenland and for a few years the escalating fishery off Newfoundland, in spite of that we were able to get back more fish simply by putting more in," Baker says. The salmon runs on the river are still not what they should be, but the water quality now is excellent. Brunswick Mining is still operating a major mine in the area, but the open-pit project is finished, all waste water now goes through a treatment plant, and the river's water quality is checked regularly.

*

The Nepisiguit Salmon Association and Baker have played an important role in the New Brunswick salmon conservation movement. Baker has been one of New Brunswick's leading salmon conservationists for two decades. In 1978, the Nepisiguit association was one of the groups pushing for the formation of an umbrella organization to speak for all the local salmon associations on New Brunswick rivers. At a 1978 meeting in St. Andrews that brought together all the New Brunswick salmon associations, a group called Save Our Salmon was formed. This group eventually became the New Brunswick Council of the Atlantic Salmon Federation, the first of six regional salmon conservation councils in North America. "We've been able to speak with one voice when we've had problems, which have been many," Baker says. Today, the Nepisiguit association has 300 members. Most are anglers from New Brunswick and other Canadian provinces and the United States, but some are conservationists who don't fish but are committed to restoring salmon runs on their river.

Robert Baker with a fish on his line at Island Pool on the Nepisiguit River.

Courtesy of Robert Baker

In *Six Salmon Rivers and Another*, Dr. George Frederick Clarke writes: "To return to a well-remembered river is like going back to the place of one's birth." Baker says of the Nepisiguit: "To those of us who fish it, it is a fantastic river. I have fished all over New Brunswick and I've fished in Newfoundland and Nova Scotia, and to me, it's the nicest river I've ever been on."

Power to the people

Before Premier René Lévesque's great reforms, Quebec's salmon rivers were the private domain of the rich and powerful. For more than 100 years, private clubs controlled all hunting and fishing grounds in Quebec. What land the clubs didn't own outright they leased from the province. The government of Quebec had given these clubs the exclusive right to hunt and fish in exchange for the protection they could offer to the resources.

The clubs were protecting wildlife, but in the 1970s the public demanded change. Quebec residents who wanted the right to fish and hunt made a strong case. Salmon fishermen were either very rich or they were poachers; illegal fishing and hunting was getting out of control, and poachers were selling wild salmon openly out of the backs of trucks on the Gaspé Peninsula.

In 1978, Lévesque's government developed a plan to open up Quebec's rivers and woods. The government created local, non-profit, self-financing community management groups called *zones d'exploitation controlée*, or, more commonly, "ZECs." These ZECs have created a progressive system of managing public resources, in which the control of rivers has been placed in the hands of communities. "They've developed a sense of ownership and pride," says Jean-François Davignon, the Atlantic Salmon Federation's regional co-ordinator for Quebec. "This is the closest thing to river-by-river management that exists."

The first ZECs were set up for hunting areas and trout lakes, and they moved onto salmon rivers by 1980. There are about 220 Quebec rivers controlled by ZECs now; about 35 rivers are run by some other type of local organization. Some of the private clubs closed down, but a good number of them stayed on the rivers. The government never expropriated any private land. Many of the private clubs had owned portions of private land and leased the rest of the river. The leases were turned over to the ZECs, and on many rivers the ZECs and private clubs now work together. On the Moisie River on Quebec's north shore, once the exclusive fishing paradise of the famous Moisie Salmon Club, there is now a ZEC working with four outfitters and the renamed Moisie Salmon Camp. The ZECs have also been peacemakers. On the Escoumins River on Quebec's north shore, natives and non-natives recently joined together to form a ZEC — and in the process ended years of feuding between non-natives and Montagnais in the town of Escoumins.

The ZECs opened up the rivers to the public, but not to free fishing. Everyone who fishes for salmon in Quebec pays a daily rod fee, and all the money goes back into the local management group and the river. As a result, poaching has been reduced dramatically. The ZECs have brought economic benefits to the river communities, creating work for guides and other people in the tourism business.

"You can't have business if you don't have fish. If you give an economic value to fish, that's the best way to protect it," says Davignon. "Fifteen years ago these guys were poaching rivers. Now they're managing them." With a ZEC, money paid for fishing is reinvested in the river communities, instead of just going into the pockets of a wealthy camp owner, or into the general government revenue pool.

Dr. Alex Bielak knows the Quebec rivers inside and out. While working on his doctoral thesis for the University of Waterloo on salmon stocks in Quebec's north shore rivers, he spent weeks at the Moise Salmon Club studying fish catches and old club records. "The tremendous success in Quebec has been the ability to get those clubs which have the riparian rights to accept a new partnership with the ZECs and with the public in general," he says. "The more people feel a responsibility toward the resource, the better management and fishing we're going to have."

✻

The Bonaventure River on the Gaspé Peninsula is a perfect example of how the ZEC system can work to manage a salmon river and create opportunities in the local economy. Glenn Legrand had been making his living guiding fishermen on the Bonaventure River for 15 years, and his fiancée, Donna Willett, had been working for former Newfoundland premier Frank Moores at his fishing lodge on the Grande Cascapédia River. They decided it was time to go into business for themselves. In the spring of 1995 they opened Camp Bonaventure, an immaculate, modern chalet on the bank of the river.

They are part of a new generation of outfitters in Quebec who are working within this new river management system. Their new lodge uses ZEC waters and also leases several private pools for its guests. The river is a mixture of public and private fishing. The Canadian Club, one of the original clubs on the Bonaventure, still operates, but the other private clubs left when the ZEC came in and took over the leased waters. The ZEC, which is run by a group of elected members, employs about 30 people and provides a protection service for the whole river. The Canadian Club employs 28 people, and Camp Bonaventure has seven employees.

Salmon runs on the Bonaventure have been holding steady at about 4,000 fish a year. The river has its own specific management plan. At the end of July every year, two divers swim the length of the river counting fish. If there are fewer than 1,000 fish in the river, the ZEC immediately rules that anglers have to catch and release fish until the end of the season. Under normal conditions, anglers on the river can keep up to seven large salmon or grilse a year.

Legrand contrasts the ZEC system with the system he experienced in 1994 on Cape Breton's Margaree River, a river that is completely open to the public. When he and a friend were fishing Cemetery Pool, there were 20 fishermen in front of them in a crowded rotation, and they were the only two Canadians in the group. The rest were American salmon fishermen who were living in campers, catching Nova Scotia salmon, putting little into the economy and nothing into the river. "This is a great system because all the money goes back into the river," Legrand says of the ZECs. "It protects the river for the future."

✻

In his report on the state of Canada's freshwater fisheries, Dr. Peter Pearse praised the ZECs as a progressive management system for the country's natural resources. He writes: "ZECs are not a case of alienating public property into private hands; on the contrary, they were established to reverse that situation and give the

Quebec public reasonable access to fresh-water fishing. They charge annual membership dues and daily access fees for fishing in their local areas to raise money to conserve and enhance the resource."

Pearse suggests that commercial sport-fishing enterprises could play a useful role in fisheries management. A number of outfitters in Quebec hold exclusive fishing, hunting and trapping rights on land leased from the province. The outfitter must pay an annual fee, hire conservation officers, submit reports and information to government and impose regulations at least as stringent as the general provincial regulations.

"Quebec's array of fishing arrangements still has many of the features I advocate in this report," Pearse writes. "It provides a wide variety of fishing opportunities, from open access to high-quality, controlled-entry fisheries. It ensures that fishing will be available to everyone for a reasonable fee. It provides a large measure of local and private involvement in fisheries management. It shifts the cost of management to the users and licence-holders, and at the same time makes many of their fisheries profitable enterprises."

Anglers on the beautiful Matapédia River on the Gaspé Peninsula, Quebec.
ATLANTIC SALMON FEDERATION

✻

On Quebec's Matapédia and Patapédia rivers, a new form of local river management has arisen that offers even more opportunities to create jobs in rural communities. Richard Firth is the director-general of a non-profit organization set up in 1992 called the Corporation de Gestion des Rivières Matapédia et Patapédia. Up until that time, the rivers had been operated by a Quebec Crown corporation that managed wildlife in the province. Beginning in 1990, representatives of the Crown corporation travelled through the Matapédia valley, asking people if the management of the rivers should be turned over to a local organization. The 23 municipalities in the valley voted unanimously to ask the government to turn over the management of the rivers to them. In 1992, the province transferred to Firth's new corporation the management responsibility for those two rivers and the Quebec portion of the Restigouche watershed. Firth's group decided not to create a ZEC, because ZECs can only charge daily rod fees; Firth's corporation wanted to offer guides, boats and accommodations as well. In 1994, the corporation brought in about $500,000, all of which went back into conservation, enhancement programs and river protection.

Glenn Legrand releases a salmon back into the Bonaventure River on the Gaspé Peninsula, Quebec.
ATLANTIC SALMON FEDERATION

The government is us

I met my river guide at 6 a.m. on a July morning at a roadside diner near the town of New Richmond on the Gaspé Peninsula. Terry Barter, 23, had sunglasses dangling from his neck and a styrofoam cup of coffee in each hand. He handed me one of the coffees and climbed into my car, and we set out to fish Canada's vice-regal river.

For an hour we drove along the road that leads upstream along the Grande Cascapédia, salmon waters that were once the private property of the governors-general of Canada and that later passed into the hands of a syndicate of American millionaires. The governors-general used to pay the Quebec government $500 a year to lease the river.

Today, the river belongs to Barter and the people who live in the communities along it. His father and mother are river guides also: Terry Barter Sr. and his wife Lina work together in the same canoe at the camp owned by the family of African gold merchant Charlie Engelhard. Members of the Barter family have been guiding anglers on this river since the 1800s. The stories of the lords and ladies, the millionaires and the Barters run together as this river flows down from the Chic-Choc Mountains, over wild rapids and through deep pools right in the centre of the Gaspé Peninsula.

Each pool along the river is numbered, part of the strictly controlled pay-as-you-fish system run by the Cascapédia Society, an organization of natives and non-natives that now holds the river lease. We drove off the pavement onto rough logging roads and eventually parked near Pool 82, a wading pool. We lugged our gear down a long path to a small screen hut by the river. Barter showed me the pool, which is a run of deep water that swirls around large boulders combined with a shallow, flat run below. He selected my flies and I fished the pool all morning as the sunlight sparkled on the water. I knew I was wading through waters laced with history.

In 1878, the Marquis of Lorne was appointed Canada's governor-general. Lord

Lorne, a poet, naturalist and salmon angler, and his wife, Princess Louise, the fourth daughter of Queen Victoria, loved the Grande Cascapédia. Lord Lorne made a deal to lease the whole river, with the exception of a few pools owned by people who lived there. In 1878, on a hill above a wide meadow that borders the river, he built Lorne Cottage, the first pre-fabricated home constructed in Canada. It was made in Ottawa in pieces, sent to the mouth of the Cascapédia on a steamer, floated up the river and assembled. Lorne called it Princess Louise Cottage. Each year they took a steamer to the mouth of the river, where they met Micmac guides who poled them upstream in canoes made from spruce and birchbark.

Lorne Cottage is now owned by the Engelhard family. Because of Charlie Engelhard's high-pressure lifestyle, his five daughters saw their parents more on the river than they did at home, and their best childhood memories are from their river; they consider Lorne Cottage home. In a privately published book entitled *Indian Summer*, Susan Englehard O'Connor writes of her love for the river. She dedicates the book to her mother, who "bestowed upon us the grand house of Cascapédia, Lorne Cottage, and its adjacent lands and waters. It is impossible for me to think of this river without a salmon breaking the surface, or to dream of these fish without the Grande Cascapédia."

O'Connor repeats a story told along the river about a special relationship Princess Louise developed with a salmon guide named Richard Duthie. Princess Louise was particularly fond of Duthie, and when she got tired of salmon fishing she would bring him up to the cottage to the "Princess Terrace," where she painted pictures of him in the nude. If it is true that Queen Victoria was a prude, it appears that her fourth daughter most certainly was not. Princess Louise, so the story goes, later set up Duthie on a ranch in Alberta.

In 1883, the Marquis of Lansdowne became governor-general. A salmon angler who fished Scotland's River Tay, Lord Lansdowne made arrangements before he came to Canada to acquire the Grande Cascapédia lease. He built a magnificent lodge beside the Cascapédia that he named New Derreen, after his home in southern Ireland. Lord Stanley succeeded Lord Lansdowne as governor-general and master of the Grande Cascapédia; he built his lodge at the mouth of the river because Lady Stanley couldn't stand the flies further inland.

The governors-general lost the lease in 1893, when Lord Aberdeen became governor-general and allowed a group of Americans to take it over. The Americans paid the government $6,125 for the river and bought New Derreen to use as a base camp. The early members of the Grande Cascapédia Salmon Club included William K. Vanderbilt, John S. Kennedy, and R.G. Dun, the founder of Dun and

Bradstreet. The club members poled 30 miles upriver to two other lodges — Middle Camp and Tracadie.

*

The Engelhard camp today has six boats, all staffed with first-rate guides, including the Barter husband-and-wife team. Terry Barter Sr. is a champion fly caster, hot-rod builder, artist, blues guitarist and writer. His wife, Lina, a Micmac, was one of the first female fishing guides in North America when she started guiding full-time 15 years ago. She handles the bow and the anchor and he runs the motor that propels the 26-foot Chestnut river canoe. The Cascapédia is a swift, dangerous river; it takes two people to operate the boat if an angler gets a big fish on the line. "I wouldn't want to fish with anybody else but Terry," Lina Barter says. "We have signs for each other. We look at each other and we don't have to speak."

They met on the river when she was at a guiding school; Barter was giving the course on how to manoeuvre a canoe. Jokingly, he asked her to marry him, and five years later, she did. When the river guests are not experienced anglers, the guides do a lot of the fishing. These two make a good team — she is right-handed and he is left-handed, and between the two of them they can cast in any conditions.

On a rainy evening in the summer of 1995, I watched them at work on a Cascapédia salmon pool named Big Curly, where a narrow rapids wraps its way through a gorge and curls around a cliff into a deep holding pool. That evening there were at least 20 big salmon in the pool. The Barters had two teenagers with them from the Engelhard camp, and they were trying to hook a salmon for them. They beached the canoe and walked along the rocks beside the pool. Terry climbed the cliff and looked down into the water while Lina (the situation called for a right-hander) cast dry flies over the fish. "Another foot out, Mom," he would call from the cliff. "You're on his tail. Now you're past him. You moved one. Put it back there again." And so on. They didn't hook a fish that evening, and when the rain really started to pelt down, they loaded into the canoe and headed back to the camp.

Terry Barter Sr. enjoys the people he meets at the Engelhard camp. For example he has talked about music and performing with jazz legend Benny Goodman and has discussed auto racing with actor Paul Newman. One salmon outing earned the

Lorne Cottage, once the Grande Cascapédia retreat of Lord Lorne and Princess Louise, is now owned by the Engelhard family. PHILIP LEE

husband-and-wife team a trip to New York City. They were guiding Jim Wolfenson, the director of Carnegie Hall. They had just finished guiding a guest who was a plastic surgeon, and Barter had been joking, "If I get you a 60-pound salmon, will you give me a free hair transplant?" They told Wolfenson about that conversation, and then they helped him hook a big salmon. Lina Barter recalls: "I said, 'Mr. Wolfenson, how big of a salmon would it take to get tickets to see Pavarotti?' He said, 'You've got your tickets.' Then when we got to the dock he said, 'I'll do better than that.' He said, 'I'll fly you and Terry down to New York for a weekend and you can do whatever you want.'" They took him up on the offer.

Terry Barter Jr. started guiding when he was 14. From the day he saw his father help one of his guests hook a big fish, he knew what he wanted to do. "I was hooked. When I'm not guiding I can't find anything to do with my time."

I spent the morning on the Grande Cascapédia in a peaceful setting without catching a fish. No matter. Terry Barter looked for salmon from up on the bank, and we chatted about fishing, writing, books, music and life on the river. Barter is a quiet, considerate man who knows how to put you onto fish. But when the fishing is slow because of hot weather and low water conditions, the days are long and hard for a guide who is constantly dealing with frustrated anglers. I wasn't one of them. I was just happy to be there and to have the chance to wet my line in those storied waters.

*

On the surface, there's no reason to think you would find harmony in the culture of the Grande Cascapédia River. In the very heart of Quebec's francophone Gaspé Peninsula, the people speak English on the riverbanks and invoke their Scottish roots by playing bagpipes to open the Atlantic salmon season. A Micmac band of 500 people lives at the mouth of the river where they net salmon. Upstream, some of the richest people in the world operate exclusive salmon lodges.

This unlikely group of people has been brought together by the Grande Cascapédia's Atlantic salmon, the famous, unusually large genetic strain that returns every year to deep pools on the spectacular mountain river. For a decade, the people of the Grande Cascapédia have shown how they can work together to protect a river, make money and create jobs.

"This is the finest salmon river in the world, bar none," says J.A. (Buddy) Campbell as he settles down in his living room to talk about his favourite subject. Across the room, Campbell's friend Bernard Jerome, a soft-spoken former chief of the Gesgapegiag Micmac Nation, nods his head in agreement.

There's a lot of history in this room. Campbell, 71, who represents the fifth generation of Campbells on the river, manages two of the exclusive Cascapédia fishing lodges. Jerome is a 48-year-old native leader whose great-grandfather worked on these waters. His grandfather was killed during a log drive on the river. Both Campbell and Jerome were instrumental in the negotiations that in 1982 led to the formation of the Société de Gestion du Saumon de la Rivière Grande Cascapédia. This organization, known now simply as "the society," manages the Cascapédia. It is run by a board of directors, half natives and half non-natives. For more than 10 years, there has been peace on the river. The salmon runs are improving and jobs have been created. The society, a non-profit organization that sells salmon fishing and a guiding service, pays the guides, hires river wardens and runs an office in the town of Cascapédia.

The Grande Cascapédia has about 96 kilometres of water for salmon angling, and fishing on this water is not cheap. Most anglers fish from a canoe, and the cost for a day on the river, which includes two guides and a canoe, is $450. Anglers have the option of sharing one rod and splitting the cost. Several wading pools cost $80 a day. The society brings in about $500,000 during the three-month fishing season and is self-financing — the only river in Quebec that pays its own way. The private clubs contribute more than $2-million to the local economy during the fishing season. "We don't want money from the government," Campbell says. "The minute you accept money, they want control."

When the society was formed, the natives agreed to restrict their netting of salmon severely in exchange for half of the jobs that would be created on the river. "Some natives were against it," Jerome says. "Some non-natives were against it. But we went ahead and did it. In our case we managed to get employment out of it, community development, and at the same time we are exercising our native rights. We continue to fish with nets, but with a very strict quota to make sure that conservation is always being maintained, because without us doing our part in conservation it would be very difficult for the Cascapédia River to handle the pressure. It was a sensible move, as far as I'm concerned."

*

The society was formed out of a concern for the future of the river. In the early 1970s, when the salmon runs began to fail, Campbell started his own salmon association in opposition to the Montreal-based Atlantic Salmon Association. The Montreal group, he says, was just a social club. "They weren't doing anything. They didn't want to rock the boat, to get the government angry. To hell with the government. The government is us."

In 1975, his association produced a report about managing the river, the recommendations of which in the end were very close to the system that was finally adopted. One of the recommendations was to stop the Indians from netting.

Buddy Campbell plays a big fish on the Grande Cascapédia River.
Courtesy of Buddy Campbell

"Three or four years later I realized we weren't going to stop them," Campbell says. By the late 1970s, a new generation of young, militant Micmac were demanding the right to fish the river. "We knew pretty well what we wanted," Jerome says. "All we wanted was part of the action. Our unemployment rate was very high and people wanted jobs. We had to do something."

At the same time, the Quebec government was taking steps to turn rivers over to the new local management groups, the ZECs. Campbell and his group had their own plans for the Cascapédia. Every river community is different and must develop its own management regime, he says. It took about a year of negotiations to create the society. "The system we came up with for this river was worked out here in this area, in this valley, and then submitted to the government," Campbell says. "The government was so damned glad to stop another Micmac war like they had on the Restigouche that they would have agreed to anything. The best ally we've got is the native people. We need them a hell of a lot more than they need us, because the governments are afraid of native people right now."

The private clubs on the Cascapédia had been leasing most of the river from the Crown. The society took over the lease. "There was no question the clubs were very happy to get this agreement," Campbell says. "They are better off now than they were before, because they don't have to organize protection, they don't have to organize the roads or the access to the pools. The society does all this. They still have the same quality of fishing. You can ask any member of the club, on this river. We refer to the public boat in the camps as 'the ghost canoe.' It interferes in no way at all."

*

Jerome says his band is exercising its aboriginal and treaty rights as much as any band that insists it must operate a major gill-net fishery on a salmon river. "I think the problem with some of the native groups is that they want to hold on to their rights as much as possible and they want to exercise them in their fashion. But in my case, if I can make a dollar out of any source of my right I certainly would go towards that end. I'm not giving anything up, I'm just using it in a more contemporary style as opposed to a traditional style.

"In our case, a good percentage of our people would rather see something like this because they see the value. They also have the right to practice fishing in a traditional manner. Mind you, it's limited. It's 15 days a year we're allowed to catch fish with nets. We have to save something for the salmon fishermen to catch. There's a business we're also protecting by doing that, by limiting ourselves. We fish only a certain portion of the estuary, making sure that the bigger salmon get up the river. We make sure enough salmon are getting upriver, for our industry and also for conservation."

Buddy Campbell supports the limited Micmac gill-netting season. "It's to guarantee their rights. They don't want to give them up. That is the one lever they've got. That's their one strong point. And they'd be crazy to give it up."

Jerome says one of the the major problems with the natives on other salmon rivers is that they have been excluded from the river communities. "Now we are part of the system and part of the protection of the species, and we feel that we

The Grand Cascapédia River is a model of co-operative, individual river management.
ATLANTIC SALMON FEDERATION

are obligated to do that because we are obligated to maintain jobs. If [people on other rivers] were to involve native people in the management of their rivers, maybe things would change a lot, giving them a chance to prove themselves."

The new society agreed that no one would be fired to make room for native employees. Instead, it would gradually build up native employment by attrition. The natives have about 30 jobs now in the outfitting business. The workforce is not half native yet, but it is moving in that direction. "Just because we're natives, we're not automatically guides," Bernard Jerome points out.

Now the native and non-native good will has extended beyond the river. "At one time the natives had their own group and the non-natives had theirs, but ever since we have been involved on the Cascapédia we have understood more of each other," Jerome says. "We participate in sports that have mixed teams. We have developed a good relationship between our community and the communities of Cascapédia and St-Jules." The reserve now has its own first-rate recreational centre. Campbell says: "As I've heard Bernard say in the past, 'There's more than one way to get money from a salmon.'"

*

Because of its beauty and its large fish, which grow to be more than 50 pounds, the Grande Cascapédia has always been one of the world's most prized salmon rivers. Five lodges now belong to the Cascapédia Club: Lorne Cottage, the estate of the Engelhard family; Tree Island, the camp of former Newfoundland premier Frank Moores; Tracadie, owned by New York multimillionaire Francis Goelet; Middle Camp, owned by Roger Baikie; and New Derreen, which is now a fishing club. In the 1930s, the Cascapédia Club built and operated its own hatchery on the river. A mounted 36-pound Grande Cascapédia salmon hangs over the fireplace inside the Angler's Club on Broad Street in New York City.

The Engelhard family has the largest operation on the river, spending at least $1-million a season there. Charlie Engelhard's five daughters — Annette, Susan, Sophie, Charlene and Sally — all have an interest in fishing and each spends two weeks a year at Lorne Cottage. Ian Fleming, the author of the James Bond thrillers, often used to come to Lorne Cottage to fish for salmon. His novel *Goldfinger* is based on the life of his friend Engelhard, who earned his fortune in gold, diamonds, rubies and sapphires.

Engelhard was a fine angler and a salmon conservationist whose portrait hangs at the headquarters of the Atlantic Salmon Federation in St. Andrews. He often stayed at Lorne Cottage and fished the river for the whole summer.

Annette Engelhard is married to the famous New York fashion designer Oscar de la Renta. When she comes to the camp, guests from the cream of New York

City society often come with her. In the summer of 1995, actor Paul Newman and his wife, actress Joanne Woodward, spent a week on the river. Most of the Cascapédia camps host celebrities. Over the years, Benny Goodman, Bing Crosby, Henry Kissinger, Jean Belliveau, Jack Hemingway, and diamond merchant Harry Oppenheimer (the director of the De Beers Group and a close friend of Engelhard) all fished the Cascapédia. Hockey great Bobby Orr fishes for two or three weeks every summer at Baikie's Middle Camp.

*

The society is heavily involved in the management of the salmon runs. All the decisions are made right on the river. In 1994, in high-water conditions, the society opened a major tributary, the Salmon Branch, for a limited amount of fishing. In 1995, in low water, the society closed the branch to create a sanctuary in pools that were filled with fish. Wardens guarded those pools 24 hours a day. Last year, anglers caught 1,100 fish with an average weight of 17 pounds. Quebec allows its fishermen to keep one large salmon a day, and seven during the season. The angler who catches a salmon is finished fishing for the day, even if the fish is released back into the river.

*

The Cascapédia camps, beautiful, elaborate lodges, share the river with public anglers, but the pools directly in front of the lodges remain private. "This was to avoid any criticism that we were giving the private camps an edge over the public fishing," Campbell says. "No way. We devised that water so that the public canoe would be over salmon at least once and sometimes twice that day. It's proven that they were catching more than the private camps."

One of the criticisms of the Cascapédia system is that the fishing is too expensive for the average angler. Campbell says: "There's no such thing as cheap salmon fishing. If you want cheap fishing go and fish catfish or eels or trout or something. Salmon fishing is the king of sportfishing, and it's not cheap."

Campbell goes fishing every chance he gets. His home is filled with fishing books, flies, photographs and wooden cutouts of his large Cascapédia fish. Jerome is also an avid salmon angler. He fishes the Cascapédia in the same way non-native fishermen do — by applying in the winter draw and paying the fees. When he left Mr. Campbell's living room he was preparing for three days on the river — days he won back in January. "Don't call me, I'll call you," he said on his way out the door.

CHAPTER ELEVEN

Watersheds

Return to the Miramichi

The Miramichi River system could become an Atlantic salmon angling paradise that produces more fish each year than all of Iceland's fabulous salmon rivers combined. This is no pipe dream. In 1995, Iceland produced about 34,000 salmon for anglers, while the Miramichi, in a terrible, low-water year, with stocks in fair shape but nowhere near their potential, produced about 25,000 fish for rods. Thirty years ago, with a booming commercial salmon fishery in the ocean and the river system's salmon runs numbering in the hundreds of thousands, the Miramichi produced 60,000 fish for anglers.

"There were big years in the 1920s and 1930s," says William Hooper, the New Brunswick government's senior salmon biologist. "I wouldn't be a bit surprised if the river was producing a million fish." A million fish! "People have forgotten what fish abundance really was like," Hooper says. "There's a whole new generation out there who have no idea what large numbers of fish in the river can really look like or feel like or sound like, because as you know they jump and constantly thrash and roll."

Hooper is a native of northern Maine who came to New Brunswick 31 years ago. After living 20 years in Fredericton, he moved to Doaktown so he could be closer to his beloved Miramichi. "I just love this place," he says. "You know, in that 30-year time period, I still have yet to see all of the Miramichi system. I only have about 10 or 25 per cent of it to go. But there's still a couple of rivers I haven't run. It's just incredible how huge it is and how variable it is and how spectacular it is." Hooper, 52, runs more than 1,600 kilometres of New Brunswick rivers every year by canoe. He is a fine salmon angler, but he enjoys just being on the rivers more than fly fishing. He often poles down the rivers with a salmon rod and his two Irish setters in his canoe.

Hooper understands the true potential of the Miramichi. He says 1967 was the last really good salmon year on the river. New Brunswick commercial fishermen that year harvested 110,000 to 120,000 Miramichi-bound fish, the Greenland

Commercial salmon netsmen in Miramichi Bay: this fishery, now closed, once killed thousands of salmon every year. ATLANTIC SALMON FEDERATION

fishery took at least 35,000 to 50,000, and the Newfoundland fishery took 150,000. "Those were large salmon harvested," Hooper points out. "We do know there are thousands and thousands of grilse harvested as well by the commercial fishermen. They were not allowed to keep them, but nevertheless they did."

That year, about 16,000 licensed anglers on the river caught approximately 60,000 salmon, and fly fishing is known to take 10 to 15 per cent of the fish in a river. "We know that after everybody got done catching these things there must have been at least a quarter of a million left to spawn," Hooper says. Altogether, these estimates total at least 600,000 Miramichi salmon — and there were probably more. In the 1930s, in the really big years, commercial fishermen caught as many as 300,000 salmon every year in Miramichi Bay alone.

*

The main Miramichi river divides into 27 large and small salmon rivers and thousands of fingers of salmon water, with no dams and very little pollution. The multiple tributaries and feeder brooks, which offer a wide variety of fish habitats to nurture the fish throughout their life cycles, have a combined spawning area of 60 million square metres. "That's an enormous area," Hooper says. "The Restigouche, by comparison, would be about half that."

The Miramichi system has a great potential to accommodate anglers. There are 1,280 kilometres of angling water in the Miramichi system, compared to 400 kilometres on the Restigouche, 28 on the Nepisiguit and 25 on the Margaree in Nova Scotia. Each of the Miramichi's salmon rivers has distinct runs of fish that enter the river almost every day from May until November. Some branches have many different salmon runs. The Main Southwest Miramichi, for example, has separate runs in June, July, August and September. Many of the strains of early-run fish, heavily exploited over the summers in small tributaries, are now extinct. But Hooper says there are still hundreds of strains of salmon in the Miramichi, all of which spawn on or about October 21 or 22, just like clockwork.

*

New Brunswick could be making a lot more money from these valuable fish. Combine the giant Miramichi system, the half-as-large Restigouche watershed, and a restored St. John watershed, which could equal the production of the Restigouche; the total is as extensive as two Icelands at least. Then add New Brunswick's other potential gems: Digdeguash, Magaguadavic, Nepisiguit, Bartibog, Big Salmon, Tabusintac and Jacquet, just to name a few. This is a world-class resource. New Brunswick could promote itself as the premier salmon-fishing destination in the world and take a multimillion-dollar step toward economic independence. If a river offers good salmon fishing, money doesn't seem to matter to rich anglers. American fishermen currently pay as much as $5,500 (U.S.) to fish salmon for a week on Russia's Kola Peninsula, not including their travel costs. On some rivers in Iceland, U.S. anglers pay as much as $2,000 a day to fish salmon. That's the bright side of the equation.

William Hooper, New Brunswick's senior salmon biologist, spends his summers running salmon rivers with his dogs.
THOMAS R. PERO

The dark side is that this future prosperity depends on healthy runs of salmon, which the rivers don't have right now. Because of the state of New Brunswick's rivers over the past decade, there are empty bedrooms in our fishing lodges while anglers go on waiting lists to spend their money in Russia and Iceland. We need the public and political desire to improve, manage and promote this great resource, instead of trying to meet minimum spawning targets, blaming the government when they are not met, and then fighting it out to see who is going to be allowed to catch the few fish that remain in the rivers.

Once an Atlantic salmon river is healthy it will look after itself. "The remarkable thing about Atlantic salmon is that if you just give them a chance, Mother Nature will produce those continuing rewards for you at no cost to the taxpayer, forever," Hooper says. "All you have to do is not take too much. It's a politician's dream. You just put this thing in gear and away it goes. You've got something forever."

In a study conducted in 1994 for the federal government, economists at the Hickling Corporation concluded that Saskatchewan has a more lucrative sportfishing industry than all of Atlantic Canada combined. The consultants were not making this up. Saskatchewan attracts 10,000 more American anglers a year than Atlantic Canada does and has more outfitters and lodges for visiting fishermen. The study said that Atlantic Canada is missing out on at least $100-million and 6,000 jobs every year by failing to promote and manage the sportfishing industry properly. This estimate doesn't even take into account the added value of the Atlantic salmon, the number one sport fish in the world.

Atlantic Canada now attracts about 8,000 U.S. anglers a year — just one per cent of the total who come to Canada — and 6,500 Canadians from other provinces. The study said Atlantic Canada could easily attract 32,000 Americans a year and 26,000 Canadians. Every 1,000 visiting anglers create 116 jobs. "We've got to start taking this resource seriously," says Les Dominy, the director of the Atlantic Salmon Federation's education and public awareness program.

A 1990 Department of Fisheries and Oceans survey gives an idea of how many jobs and dollars can be generated on New Brunswick rivers. The survey showed that 425,000 people fished for sport that year in the Atlantic provinces, 3.5 per cent of them non-residents. These anglers spent $243-million and created 49,000 seasonal jobs. In New Brunswick, 104,000 anglers spent $61-million and created 12,000 jobs; about 22,000 salmon anglers spent a total of about $20-million. In that same federal survey, 65 per cent of the anglers said they would be willing to pay more to fish and 9,300 of them said they did volunteer work for the resource.

In a 1992 study, resource consultants LGL Limited determined that improved salmon and eastern brook trout stocks would mean another $50-million a year for the New Brunswick economy and a similar amount for Nova Scotia, Prince Edward Island and Newfoundland. Dominy says the region can support an increase in the number of anglers — but only if we practise good conservation.

He points out that the 1990 federal survey showed that 104,000 anglers in New Brunswick caught four million fish, an average of 40 fish per angler. (This includes salmon, trout, bass, mackerel, smelt and other species.) "We're still killing too many fish," Dominy says. "As most anglers realize, our fresh waters are producing fewer fish than they did several decades ago because of lower summer flows, higher water temperatures, and more pollution. For ocean-going fish like salmon, we've constructed more dams to obstruct their migration, and for reasons scientists can't explain, fewer fish are returning from the ocean. We can't continue to fish as we have in the past, catching everything in sight, killing everything we catch."

River conservation is essential for this industry to grow, and even the Miramichi, New Brunswick's bright spot, has its share of problems: deteriorated habitats; logging roads crossing streams improperly and causing siltation in spawning grounds; and polluted water in the estuary. In the Little Southwest Miramichi, the numbers of small fish are lower than they should be, and biologists are not certain what the problem is. Some anglers in the headwaters of the Miramichi are concerned that the river is being overfished because the season now extends from early spring until late fall throughout the river system.

"There have been years in the past when we've had these crashes in fish abundance," Hooper says. "And the crash we're into now started in about 1990 and is still with us, but I can't help but think we're going to pull out of that. The only thing I'm concerned about is that we do everything we can to maintain the small parr populations at as high a level in the rivers as possible. That's the only thing we can do at this point as fishery managers. The ocean is too vast, too huge, and we don't understand it well enough to do anything but be concerned about it. We just have to let it ride and say, 'This has happened before.' What we have to do is focus on our river environment to make sure that we're producing as many juveniles there as we can, so that when the break comes we'll be throwing out a lot of smolt to take advantage of the improved conditions in the ocean."

*

While the Miramichi and other giant salmon producers are important, other small rivers in New Brunswick need attention. For example, Hooper says, the Bartibog River, just above the Miramichi, is a salmon stream that most countries in the world would give anything to have. Outside New Brunswick (and even inside the province, for many people) the Bartibog is unknown.

"This is an absolutely remarkable stream for Atlantic salmon," Hooper says. "It could be a world-class stream, but we really need to get at managing it. It doesn't really belong to anybody. If something belongs to people, they cherish it, they want to look after it. If it's all 'free beer,' then it's really not appreciated. The Bartibog was always sitting there for people to go and fish and have their way with the stream and leave. What's wrong with it is basically that we haven't allowed enough fish to spawn over the years to populate it. We've had no regulations to ensure that the river would populate itself and Mother Nature would take over." Today, a new Bartibog Salmon Association is beginning to take steps to manage that river.

The province has recently turned its attention to another small northeast coast river, the Jacquet, and has put a barrier fence on the stream to hold the fish downstream and stop poaching in its headwaters. The barrier is starting to turn that river around. "We have other rivers around the province that are just fabulous salmon rivers, potentially, if they are just given the opportunity," Hooper says. "We just have to manage them, and they'll pay back in spades."

After 30 years of running rivers, which stretch does Hooper like best? "I do have favourites," he says. "I'll pass on that one, for obvious reasons. I might take you there sometime but I won't tell you."

The day has come

The Miramichi River is bridging the distance between the movers and the shaken. In the 1980s, when commercial netting threatened to destroy the river's salmon runs, Miramichi people came together to fight for their future and convinced the federal government to stop the slaughter. Today, with the salmon resource still threatened both at sea and in fresh water, and governments talking about pulling out of the salmon-management business, the people of this great river system have come together again and taken a giant step towards taking control of their watershed.

Renous native Bernard Duffy is the driving force behind this movement. He is a man who suits the character of this salmon river — "a great working river," in the words of David Adams Richards. Duffy, 41, is not a politician or a millionaire angler. He's not a very good flycaster and doesn't really care if he ever catches a salmon. He is the son of a woodsman and river guide, a welder by trade, who wants future generations to be able to enjoy these waters as much as he has.

Bernie Duffy is the driving force behind the Miramichi Watershed Management Committee. PHILIP LEE

Duffy has boundless energy and simply refused to quit when people told him that it would be impossible to organize the diverse group of people who live and work along the Miramichi. "Bernie Duffy is the closest thing to perpetual motion you're ever going to see," says Jack Fenety. In September 1995, Duffy was elected president of the first Miramichi Watershed Management Committee. This committee has great potential for the Miramichi, and its example could be followed on every watershed in Canada.

The creation of this watershed committee is no small accomplishment. Duffy has formed a group that involves all of the people with a stake in the river: the powerful Miramichi Salmon Association, which represents many riparian owners on the river; three native bands — Burnt Church, Eel Ground and Red Bank; the outfitters and the guides; all the smaller river associations along the river; Heath Steel; and forestry giants Repap, Avenor and J.D. Irving. The committee operates in the most simple way possible: Its members all sit around the table as equals, each stakeholder gets one vote, its monthly meetings are open to the public, and the future of the river is always on the agenda.

"This river is one of the greatest resources that we have," says Duffy. "We have the one river system that has lots of fish and very little pollution. If it's looked after, it's something that's always going to be there." If the watershed committee is to succeed, there must be give and take on both sides — governments must step back and river communities must step forward. "How do you get it out of the hands of politicians?" Duffy asks. "That's going to be the stickler."

The federal and provincial governments have said that they are willing to step back from the day-to-day river management. Just how far they really want to go remains to be seen. Atlantic Salmon Federation president Bill Taylor says New Brunswickers should seize the day. "Community watershed management holds the best future for our salmon rivers. We have collectively been criticizing governments for decades about how poor a job they have been doing. Now we have been presented with an opportunity to step forward, grab hold and demonstrate that we can do what the government could not do, and that is efficiently and effectively manage our rivers as we see fit." Taylor says watershed management committees should bring together river communities — people directly involved in the salmon industry and independent voices from outside the fishery.

Duffy became a river organizer in 1990 when he helped form an association on two Miramichi tributaries, the Renous and the Dungarvon. Eventually, Duffy decided it was time to organize the entire Miramichi watershed. "The way to do it was to get river management where everybody on the Miramichi would be the voice of that watershed. So that's what I undertook to do, and it turned into quite a job to say the least."

In February, 1995, he organized a meeting at the Wharf Inn in Newcastle. "I don't know how it happened, but everybody ended up with a representative at the meeting. We had quite an evening." At that meeting, the group decided to hire a professional facilitator to help them find a way to work together. The group brought in Tom Patello, a facilitator from a Nova Scotia company called Catalyst Training. At a two-day session on the Miramichi, Patello didn't let anyone argue. He broke them up into groups for discussions. "You could just see the people changing, their attitudes changing," Duffy says. "We decided then and there that we were going to try to move this thing ahead."

The big question was how to organize the group. Duffy discussed this thorny question with Alex Mills, Vincent Swazey and Bud Bird. They decided that they needed the blessing of the river's power broker, Jack Fenety, who was vacationing in Florida during all the initial planning for the watershed committee. Fenety's Miramichi Salmon Association, the most influential conservation group on the river, has hundreds of members in Canada and the United States. "Everything hinged on Jack, whether he was going to support it or not," Duffy recalls.

When Fenety returned to New Brunswick, they met in Bird's Fredericton office and told him what had been happening. Fenety agreed that river management was the way to go. How many representatives would Fenety want on the committee? Fenety answered: "We've got a thing in this country called democracy. You have one vote, I have one vote. You can't let the strong step on the weak. There's one vote regardless of who you are. That's the way it is. I won't support it any other way." That clinched the deal. Duffy says: "When people saw that, it just seemed to fall into place. In five years' time I think you will find that river management will be *the* voice that will make the decisions on whatever takes place in this fishery."

*

Duffy, as always, is optimistic, but a lack of real decision-making power has been the stumbling block on the road to watershed management in New Brunswick. The Miramichi Watershed Management Committee is not the first group to try to take control of its own river. On a smaller scale, New Brunswick's Hammond River Angling Association has been involved in watershed management since a group of conservationists banded together in 1977 to try to preserve their river, which is a tributary of the Kennebecasis.

Today, this impressive association has more than 350 members, is incorporated, has a board of directors and holds monthly meetings. Its list of accomplishments on the river is long. The Hammond River group has been a strong voice fighting against pollution in the river system, in its early days helping to stop a sewage treatment plant planned for the watershed, and, almost 20 years later, exposing a spill of brine from the Potacan mine that was killing small fish. It has also been involved in a school education program in which students have hatched salmon eggs right in the classroom and then released their fish into the river. Members of the association have been cleaning up the river every year and doing habitat work — planting trees to stabilize the riverbanks and trying to restore spawning grounds in the river's small tributaries.

The Hammond River Angling Association has also acquired 31 acres of land, donated by the Irving family, and plans to build a $100,000 conservation centre that will eventually have its own hatchery. This group has shown that it is responsible and dedicated to the conservation of the river, and its members have been at the forefront of the salmon conservation movement in New Brunswick. The Hammond River association's first president, Lou Duffley, was also the first president of the New Brunswick Salmon Council, which represents more than 10,000 volunteers on rivers around the province and has been an important conservation force. However, Ottawa has so far refused to allow the Hammond River group to make decisions about the management of the waters these volunteers have been watching over for almost two decades.

Since 1993, the Hammond has been closed to salmon angling because of low fish returns in the St. John River system. But unlike the St. John, the Hammond is meeting its minimum spawning targets. In 1994, when the association asked for a hook-and-release fishery on the Hammond, Ottawa chose instead to close the entire St. John watershed. "Regulations governing the harvesting of Atlantic salmon for the entire St. John River drainage are based on returns of fish to the Mactaquac Dam just above Fredericton," says Paul Daigle, a past president of the association. "The only problem is that the Hammond River fish don't go to Fredericton." He points out that Hammond River salmon take a right turn soon after going through the Reversing Falls in Saint John and head up the Kennebecasis to the mouth of their home river. Daigle wonders how scientists think they can predict the runs on the Hammond by counting fish at the Mactaquac Dam. He says nobody knows more about the Hammond, or cares more for its future, than the members of his association. It is frustrating for them to work so hard and then not have their advice taken seriously.

*

Anglers are going to have to start paying more to fish, Duffy says. "It costs us less than a penny per hour to fish, if you look at the season that we have now. You can't continue to have a river system with fish in it for that price." Duffy says he would like to see people still buy a fishing licence from the province but pay extra for a "conservation stamp" to fish the Miramichi. This stamp money would go back into the watershed to pay for more wardens, restocking programs and whatever else the river needs. If each angler who now fishes the Miramichi paid an extra $10 a year, the watershed committee would immediately raise $224,000 for the river.

"If the public goes to the government and says, 'Look, we think we should be paying $10 more to fish and we think that money should be going into the river for restocking,' the people have their say. The money has to go back to the river. To me, the price of a box of beer wouldn't be too much to increase the licence. You can make up your mind: 'I'm going to buy a box of beer for the weekend, or I'm going to give up one box of beer for the year and I'm going to go fishing. I'm paying so that things are going to be better in the future.' That's what you're doing — buying your future."

Duffy also thinks outfitters who use public water should put money back into the resource. Duffy himself has an outfitter's licence, although he is not actively selling fishing now. He owns a cabin and two pools on the Renous. "As an outfitter I don't think I have the right to bring in non-residents and charge them $300 a day to go out and fish public water. I can have six non-residents in my camp today, charge them $300 a day, take them down to the public water on the Renous, crowd the taxpayer out, kill 12 fish and not put one penny back in. It doesn't make any sense. It's the only business you can be in where there's big money and not one cent of investment. Why doesn't that outfitter have to pay for the use of public water? I don't own it, so why am I allowed to charge people for using it? If you own a business and you want to use the public resource, you should have to pay for it."

Duffy also thinks we need to open up more water for the public to encourage anglers to fish in the watershed. Remember, if Duffy has his way, all of these anglers would be putting money back into the river. He says there is a lot of public water open for anglers, but most of it is inaccessible. He says the committee could create river access roads and then produce maps to help visiting fishermen find their way around the river.

*

The watershed also needs more wardens guarding against illegal killing of valuable salmon. The watershed committee could hire its own river guardian staff. "We need more protection, there's no two ways about it," Duffy says. "There is a poaching problem. Our wardens, federal and provincial both, are seasonal people, yet they take their holidays in the summer. They take their training in the summer. They go to get requalified for their firearms in the summer. They service their vehicles in the summer. Why isn't this done in the winter?

"You've got six federal wardens today, somewhere from Juniper to Tabusintac. Now there's probably two of them in court, two of them are on holiday, one guy could be getting the vehicle serviced. Another guy is out on sick leave. Who's looking after the river? No one. If we're generating $30-million plus on the Miramichi watershed, doesn't it make sense to put somebody back in to protect it? If this fishery is generating the revenue which the province claims it is, why are they not prepared to put more into it? That's what I can't understand. If the user has to pay to get more protection, that's what we have to do. We can either lose it, or we can pay more and keep it."

*

Bill Taylor says the Atlantic Salmon Federation is planning to assemble a group of professionals that can provide advice to watershed management groups and help them develop a plan for their rivers. River communities should be working towards taking complete control of their watersheds, he says. This would include everything from licensing and protection to habitat restoration and stock assessments. All of this will cost money. "That means, in some cases, the organization may feel they should be charging daily rod fees and reinvesting the money that is generated back into the management and rehabilitation of the resource. In some cases it may mean just a season fee to fish that particular river."

Bernard Duffy argues that the river-by-river management system could be a major source of new jobs for New Brunswick if the provincial government pushed the program forward and launched a major salmon river preservation campaign. This conservation program would create employment for river guides, wardens, road builders, cooks, biologists and general labourers. The program would not cause pollution or require any government investment. "The day has come for river management."

CHAPTER TWELVE

Home Pool

Our rivers remain

It is not yet too late to apply remedial measures but we must act quickly else the Atlantic salmon is doomed to extinction.
— Dr. George Frederick Clarke

After you have exhausted what there is in business, politics, conviviality and so on — have found that none of these finally satisfy or permanently wear — what remains? Nature remains.
— Walt Whitman

A short drive from my home in the St. John River valley, the Nerepis River flows into the St. John at Woodman's Point, a jut of land where the big river completes its straight, wide run through Long Reach. Here it turns sharply down into Grand Bay and joins with the Kennebecasis River on its final rush into the Bay of Fundy. The headwaters of the gentle, lazy Nerepis run down from the Gagetown Armed Forces training area. Sometime in the last decade, the Nerepis River's Atlantic salmon were driven into extinction. No one knows precisely when or why this happened. All we know is that during the time when this important part of New Brunswick's natural heritage was being lost forever, no one took decisive steps to save it.

In the fall of 1995, soon after I started writing newspaper stories about Atlantic salmon conservation, residents in my community began telling me about the Nerepis River's lost salmon runs and proposing theories about what may have happened: a new causeway and series of bridges at the mouth of the river may have turned the fish away from their home waters, or heavy poaching in the headwaters may have finally beaten them into extinction. Even if we are unable to discover the precise biological cause of the loss of the Nerepis salmon, we can at least understand the practical cause of this environmental tragedy. The Nerepis salmon were lost because no one owned them, and our governments, the stewards of the resource, simply haven't done the job that we have depended on them to do.

I offer another case in point. One rainy spring Saturday, my father and I took a drive to the headwaters of the Kennebecasis River to visit one of our favourite trout-fishing haunts. When we arrived and started walking upstream in our waders, we discovered to our horror that logging crews had been at work during the winter on the banks of this pristine little stream that is dear to our hearts. Huge skidder ruts ran right alongside the water, and at one point the bank had completely caved in. The stream was spilling out of its banks into the new logging road. Someone had thrown some debris onto the bank to try to stop the flow, but it wasn't working.

This sort of reckless logging was not only destroying precious trout habitat but also endangering Atlantic salmon. Many times while fishing for trout on the upper reaches of this river, I have hooked and released salmon parr. A careless logging crew had violated a trout and salmon nursery and had done it with impunity. Every stretch of water needs our undivided attention, and the waters of the upper Kennebecasis, like the waters of the Nerepis, aren't getting it.

During the course of my journey along salmon rivers in Canada and overseas, I discovered a number of resource management systems that are working, from the enlightened ZECs and Grande Cascapédia Society in Quebec, to the intensively managed rivers of Iceland, to the forward-looking Tweed Foundation in Scotland. Here in New Brunswick we have seen the fine work of the Big Salmon River Angling Association, the Nepisiguit Salmon Association and the Miramichi Watershed Management Committee. Taking lessons from a variety of these sources, I will outline a plan that I think would protect our rivers and save our salmon. I am presenting a broad outline of a river management plan, its details deliberately vague, because each river is different, each river community is unique, and they must all be allowed to create systems that work for them. The philosophy behind the plan is a transfer of power and ownership from governments to friends of the rivers. The legal clout behind the plan is riparian rights, secure property rights that would allow river owners to turn immediately to the courts to combat water pollution, habitat destruction, dams, overfishing or any other threat to their river.

In the course of my research, I met government scientists, men and women of great intelligence and ability, whose passion and love for rivers was too often stifled by politics and bureaucracy. Invariably it was more interesting to hear what they really thought about the state of our rivers than what they were willing to say for the record. The chief biologist in charge of the St. John River watershed, and therefore the person responsible for the Nerepis and Kennebecasis salmon, now works outside of New Brunswick's borders in a government office building in Halifax, Nova Scotia, a five-hour drive away from waters that badly need the undivided attention of a fine scientist. We need to find a way to put these professionals back on the rivers.

*

Governments will not give up their powers easily, however. In the winter of 1996, when New Brunswick's Natural Resources Minister Alan Graham was working on new legislation that would give the people of this province more local management control over rivers, he told me that he was meeting with resistance from some of his bureaucrats. I expect similar resistance from some people on the rivers. I have spoken to volunteers who have been involved with angling associations on New Brunswick rivers, people who have been doing valuable conservation work. When I told them that I admire what they have done but that they were going to have to do more in the future, some of them replied: "Hold on a minute. We're not the government." I maintain that we should take seriously the words of the Grande Cascapédia's Buddy Campbell: "To hell with the government. The government is us." The people of the Grande Cascapédia River secured a long-term lease of the famous salmon river from the Quebec government and formed a self-financing, local management regime that makes sense to them.

Whereas New Brunswickers are going to have to do more to help manage the rivers, anglers who use them are going to have to pay more. As a person who has spent many fine days angling on New Brunswick's public salmon waters, I sympathize with those who are distressed by the idea that we will no longer have free fishing and unlimited access to waters. But I am willing to give up some of my fishing days in order to see this resource preserved. Most salmon anglers in Iceland save up for several good outings a year. When they do go to the rivers, the water, angling and scenery is magnificent. I'd be satisfied with that. The principle is simple: Those who use the resource will have to give something back.

*

Property rights can work for conservation on New Brunswick rivers, as we have seen on our privately owned and Crown-leased waters. Iceland's farmers protect their waters in a management system that is both good for the salmon and largely self-financing. Scotland has had privately owned rivers and private ocean netting rights for hundreds of years, a system that has its critics today and may indeed be in need of some reform but that nevertheless has done a remarkable job of protecting the resource.

Despite these examples of successful private stewardship, influential voices in the conservation movement oppose the idea of private ownership of our resources. Their views reflect, I believe, a fundamental misunderstanding of the nature of property. British essayist G.K. Chesterton articulated this misunderstanding in a 1931 essay about sex and property. Chesterton was fighting the "fashionable" critics of his day (I think today's fashionable critics fall into the same category) who thought of property only in the sense of money, in the sense of salary, "in the sense of something which is immediately consumed, enjoyed and expended; something which gives momentary pleasure and disappears."

Chesterton said that his critics didn't understand that property is something that involves incidental pleasure "but begins and ends with something far more grand and worthy and creative." He gives the example of a man who makes an orchard where there was a field. The man, who owns the orchard and decides who will inherit the land, also enjoys the taste of apples and cider. But Chesterton argues that this man is doing something "very much grander, and ultimately more gratifying, than merely eating an apple. He is imposing his will upon the world in the manner of the charter given him by the will of God; he is asserting that his soul is his own, and does not belong to the Orchard Survey Department or the Chief Trust in the apple trade. But he is also doing something which was implicit in all the most ancient religions on the Earth; in those great panoramas of pageantry and ritual that followed the order of the seasons in China or Babylonia; he is worshipping the fruitfulness of the world. The notion of narrowing property to merely enjoying money is exactly like the notion of narrowing love to merely enjoying sex."

Are there some among us who would prefer to continue to be the servants of the Orchard Survey Department, even if that department is doing a poor job of sustaining our orchards, and even if other areas of the world without Orchard Survey Departments are growing better apples?

As Chesterton points out, there are limits placed on the property owner. In his essay, "The Enemies of Property," he describes property, the family homestead, as the "art of democracy," a place where people express themselves creatively. "It means that every man should have something that he can shape in his own image, as he is shaped in the image of Heaven. But because he is not God, but only a graven image of God, his self expression must deal with limits; properly with limits that are even strict and small." God enjoys unlimited creation. People enjoy creation with limits, taking pleasure in the possession of the land, but also partly possessed by it.

In areas outside of river ownership and riparian rights, private property conservation has had a long and distinguished track record. In Great Britain, the National Trust for Places of Historic Interest or Natural Beauty is the country's largest conservation organization and largest private landowner. One of the trust's major campaigns in recent years has been to buy distinctive sections of coastline that it wants to preserve. By the end of the campaign it will own a third of the entire coastline of England, Wales and Northern Ireland. The trust buys wild spaces, cultivated gardens, farms and historic buildings, all in the name of preserving Great Britain's natural and cultural heritage. And it has extra legal clout from a 1907 act of Parliament that gave it "inalienability," a legal right that protects its lands from government seizure for public works projects. The British Parliament had the good sense to protect the trust's holdings from governments.

The United States has the remarkable example of the Nature Conservancy, a private land trust that was founded in 1951. This private scientific and educational organization, which works to preserve plants, animals and natural communities, owns or manages 7.5 million acres of ecologically sensitive land in the United States. Since 1962, the Nature Conservancy of Canada, an organization independent of its U.S. counterpart, has helped to protect 1.2 million acres of land and 600 nature preserves. It often buys land and then establishes management regimes through local conservation groups. In recent years, the Nature Conservancy in the United States has entered into land management contracts with the United States government, a system where the government retains ownership of land and the Conservancy manages it. This is the kind of system I am proposing for New Brunswick rivers, and I would invite groups like the Nature Conservancy to take part.

*

I will describe my river management plan from the top and work down, remembering that in the end it should include management of each individual river; it should also include secure property rights on our rivers held by a wide range of individuals and groups: community watershed associations, private owners, nature trusts, clubs, outfitters, angling associations — whatever sort of arrangement suits each river culture.

The first step would be to pass the responsibility for conserving New Brunswick's Atlantic salmon from the federal government in Ottawa to the provincial government in Fredericton, a city divided by the waters of the St. John, a wonderful but threatened salmon river. The provincial government would set general guidelines and priorities for this resource but would have no day-to-day management responsibilities. The minister of natural resources would develop the legislative framework to allow a local river management system to work. Local river managers could apply for long-term leases to all Crown-owned waters. The minister would also appoint a new key player in this plan: an auditor-general of the environment.

The auditor-general of the environment would be independent of government and would report directly to the people about the state of our salmon rivers and the state of our environment generally. The auditor-general would have the power to terminate a river lease if a local manager is threatening the resource. One of the criticisms of Scotland's river management system, a regime that is entirely based on property rights and local management, is that there is no effective independent general overseer of the resource; appointing an auditor-general of the environment would address this problem here. The auditor-general's office would have its own team of scientists, who would assess fishing plans and habitat-restoration programs in ways similar to those of Iceland's Institute of Fisheries and Aquaculture Research. This system would remove politics from fisheries science.

Immediate steps need to be taken to try to avert disaster on a series of severely threatened New Brunswick rivers. The provincial government should sell Larry's Gulch, its exclusive, money-losing fishing camp on the Restigouche River, and use the million or so dollars it raises from the sale to finance an independent scientific task force to tackle the pressing environmental concerns, especially in the Bay of Fundy rivers like the St. John and Big Salmon, where salmon runs are in serious trouble. Instead of using Larry's Gulch to entertain its special business guests, the province should spread its economic development dollars among a variety of professional New Brunswick salmon outfitters who have top-quality camps.

River associations could model themselves either after the Miramichi Watershed Management Committee, where all the players in the community — from rich Americans to river guides — are equals and get one vote, or after the simpler structure of the effective and admirable Big Salmon River Angling Association, which has a solid track record of doing what it thinks is right and getting the job done. These river associations could be given long-term leases for all Crown-owned waters.

On those rivers where associations can make money selling salmon angling, the associations would pay fees to help support the office of the Auditor-General of the Environment and its team of scientists. Eventually, with improved salmon runs, the office would be financed entirely from these river associations.

In some cases, river associations may not be the appropriate bodies to manage a watershed. For example, the lower St. John River is badly in need of a steward with secure riparian rights, but that stretch of river will never be a money-maker from salmon fishing because all the angling takes place further upstream. A conservation land trust like the Nature Conservancy would be the appropriate body to step in and fill the gap there.

The friends of the rivers would make their own rules, subject to review by the auditor-general, and subject to one legislative restriction: The New Brunswick government should maintain the strict catch-and-release policy for large salmon in perpetuity. (Anglers should be able to continue to kill grilse in rivers where there is a surplus of fish.) The river associations could charge daily rod fees, issue season licenses or sell fishing by lottery. The provincial government would no longer issue general licenses; these would be obtained on each river. The associations would have to raise money to hire wardens, biologists, habitat experts, cleanup crews and road-maintenance crews; this program would be a generator of jobs in rural communities. However, associations would have to make the people who come to their rivers pay for fishing, and all money raised from angling would go directly back into river conservation. The private camps and outfitters would continue to operate just as they have been operating, although property taxes on salmon waters and lease payments would go directly into the river association's budget.

Fishing seasons would be determined by conditions on the river. There would be restricted fishing in low, warm water conditions because too many valuable salmon die when hooked and released in warm water. Fishing in the cool waters of the spring and fall would be less restricted. As soon as fish are seen digging spawning nests, the auditor-general could close a river if the river association hasn't already closed it. Every river would maintain sanctuary areas for spawning fish, areas never violated by anglers.

The province would take steps to ban gill nets from its waters. If this ban becomes a problem legally because of aboriginal and treaty rights, I would encourage native bands to lift their nets voluntarily. I would also encourage partnerships on the rivers — the Grande Cascapédia is an excellent model. These partnerships would offer native bands secure, long-term employment in the angling industry as an alternative to netting these valuable fish. I would encourage native bands and the federal government to sit down and negotiate new treaties that make sense in today's world and don't require numerous trips to the courtroom to sort out the meaning of ambiguous, dated language.

*

One of the immediate challenges for this plan is finding a source of money. Many rivers are far from the point where they can be self-financing, and some of our smaller rivers may never be self-financing. Perhaps Iceland can provide a solution to this problem. In Iceland, hydroelectric projects all pay to help support salmon river management and scientific research. However, here in New Brunswick, N.B. Power pays nothing for the use of the St. John River water, from which its hydroelectric dams draw enormous revenues. By comparison, in 1994-1995, N.B. Power paid $142-million for coal and fuel oil to feed its thermal generators, which produce slightly more than twice the power of the dams.

Although N.B. Power's multimillion-dollar hatchery at the base of the Mactaquac Dam is important, it is now time to face the fact that the hatchery doesn't adequately compensate for the demise of the legendary salmon runs on New Brunswick's largest river. If N.B. Power were charged, say, $20-million a year for using the St. John River water, a fraction of the cost of the fuel for thermal generators, this amount would give river associations and the auditor-general of the environment a steady cash flow. This levy could mean a slight increase in power rates (N.B. Power's sales revenues now approach $1-billion a year), but it would be a legitimate cost, not a charitable donation. It would compensate New Brunswick for the use of a public resource and for the environmental damage it has caused. And New Brunswickers, especially industries that would pay any increase, should accept honest pricing of electricity to protect one of New Brunswicks most valuable cultural and economic resources.

The provincial government should require N.B. Power to turn off its dam turbines for short periods of time in the spring and fall to allow smolt hatched in the headwaters of the river system to migrate safely downstream. The province's new salmon task force (financed from the sale of Larry's Gulch) could investigate the timing of smolt migrations to minimize the cost of such a program, but finding a way to get the smolt to the sea alive is crucial for the future of the St. John River salmon.

*

We must also turn our attention to the enormous environmental problems in the sea, for they are perhaps the greatest threat to the ultimate survival of the salmon. In the first place, the federal government should create no-harvest zones: at least 25 per cent of a fish stock's range should be left to lie fallow, ensuring that even in the event that there is overfishing, a strong remnant population remains. There should be no fishing on the ocean spawning grounds. Spawning and nursery areas must be protected if ocean stocks are going to survive in the future.

Second, limited offshore fishing rights should be sold at auction as secure, transferable quotas. The money from auctioning off these quotas could be used to establish an independent ocean scientific trust fund to perform scientific studies on ocean fish stocks.

Next, it is essential that ocean fisheries be restricted in their use of technology. In Scotland, there have been restrictions on technology for fishing salmon at sea for hundreds of years. There should also be a permanent ban on industrial fishing and the killing of small fish at the lower end of the food chain. The capelin would then be preserved so that they would again run freely onto the Newfoundland beaches to spawn — and thus create food for larger fish.

In addition, communities along the coast should be given secure property rights — riparian rights — to a section of coastline and a fishery. These communities then could work out a sustainable fishery plan that makes sense to them, and they could be given responsibility for managing their own section of water.

Furthermore, fish farms should be kept away from salmon rivers and, as soon as the technology is available, moved to inland facilities. We don't know whether farmed salmon are harming wild salmon runs, but we know that it is a possibility. By the time we find out for sure, it may be too late.

The federal government should permanently close the Newfoundland and Labrador ocean salmon net fishery, and all river associations managing fish stocks that migrate into the North Atlantic should contribute money to Orri Vigfusson's North Atlantic Salmon Fund to help him buy out the Greenland and Faroe Islands' high-seas salmon fisheries. These measures might lead to permanent solutions to the high-seas slaughter.

*

In the end, we must be driven by a desire to find a way to live in balance with nature, to seek quiet, pure rivers filled with fish. At a certain point, our lives and our relationships with our rivers have to make sense. Right now they don't. This is a call to action in the great hope that the wars on our rivers will end and salmon will return to our waters in great abundance. After we have exhausted what there is in business and politics, our rivers remain. They must be allowed to go on living.

The river of life

One autumn night in your thirties
you bob up from sleep.
You've been dreaming of rivers.
Dazed, warm in the blankets,
you turn the dream over,
burnish the edges, hidden corners,
before the glow goes flat
in the morning light.

— Eric Trethewey

I've been dreaming of rivers since the soft summer evening when my father and I turned right at the Robinsonville bridge and drove along the road that follows the Upsalquitch River upstream. I came to the conclusion some time ago that this search for a way to save our salmon rivers is essentially a search for a better way to live. During my search, I asked people to tell me their river stories. Then I told you what they told me. I have one more story to tell you, and it begins when my father and I turned right at the Robinsonville bridge.

The road that follows the Upsalquitch River upstream is at times dark, almost covered by a canopy of outstretched hardwood branches. Then you break into the open light, and on your left you can see a band of rippling water. Through one of these breaks in the trees you can see the Home Pool, part of Tucker Cluett's Crown lease, which for me is the most beautiful salmon pool in the world.

That evening, my father and I parked the car and looked down on the scene before us. A canoe was anchored in the middle of the Home Pool, and a man was standing up in the boat, casting. River guide Shane Mann sat in the stern of the canoe, watching the water. On the beach, my father's friend Leon Hatch stood smoking his pipe, leaning forward slightly toward the river, with his two dogs circling around his feet. No one said a word. We were all waiting for a salmon to rise.

My father and I drove on until we came to the bridge farther upstream, where we crossed the river and headed back downstream on a dirt road. When we arrived at Hatch's farm, we turned toward the river and drove down through his pastures, the tall grasses scraping the bottom of the car, to the old camp that was once a salmon club. If you walk down the front lawn of the camp and on to the beach of the river, you're looking at the Home Pool.

*

Home Pool on the Upsalquitch River. PHILIP LEE

The camp, which has recently been purchased by the Hatches, was owned by one of my relatives, Arthur Howe, who lives in Connecticut. He is married to my grandmother's first cousin and is a great outdoorsman and salmon fisherman.

I had travelled to the Upsalquitch to work. I had plans to conduct interviews up and down the river, visit famous Restigouche fishing lodges and research court records in Campbellton. My father was with me for a few days to keep me company and do a little fly fishing on the side.

The old Upsalquitch camp is a simple building — a living room with a fireplace, a bedroom, a kitchen and a screen porch. From the screen porch you can see the river. There used to be a guest house and guide's house on the property, but they were swept away in a spring flood. In the evenings, we walked down to the river and watched the guests and guides fish the Home Pool. The anglers were either from Cluett's MacLennan Lodge or Joe Sewall and Richard Warren's Millbrook Farm. These lodge owners are all from Maine. The Home Pool and the waters that are now leased to Cluett used to be leased to the old club. My family has been staying at the Upsalquitch camp for a couple of weeks most summers for almost two decades. In recent years, with Cluett's permission, my father has fished for salmon in the Home Pool in August (the lodge guests fish only during the river's best six weeks, in June and July).

Inside the old camp are stories of the river. Some of them are written on the wall boards; others are recorded carefully in books. During the years we've been visiting this camp, these fishing records have become part of our family history. Old fishing records — and there are log books like these on salmon rivers all around the world — tell the story of our rivers. Scientists realize these sorts of records are a valuable tool in understanding salmon biology. Dr. Alex Bielak studied fishing logs at angling clubs on Quebec's north shore when he was researching his doctoral thesis for the University of Waterloo. My reading of the Upsalquitch log books is entirely unscientific, but not, I think, without substance.

The early records written on the wall of the Upsalquitch camp are evidence of the great fish runs we no longer have. And despite the fact that people keep assuring us that we have had ups and downs in the past, fishing from 1915 until 1935 was fairly consistent. For example, in 1915, two rods landed 87 fish in 26 fishing days. In 1925, two rods landed 104 fish in 29 days. In 1935, two rods landed 54 fish during 11 days.

*

The 1940s were really good years. One trip in particular stands out. On June 21, 1942, E.S. Sherwood and M.M. Peterson arrived on the train. They lost their baggage but made it to the camp safely. The water was swift and high, 52° F. (Beginning in 1915, all the anglers were meticulous about taking water temperatures.) On June 22, Peterson, who had never hooked a salmon before, and his guide, a man named Harris, fished the lower run. There, Peterson killed his first two salmon — one 9 pounds and one 11 pounds. Sherwood, meanwhile, was having the most exciting day of his fishing experience. He landed three salmon at the Home Pool and had a total of seven fish during the day. That day, 21 bomber planes flew over the river.

On June 23, Peterson and Harris fished at a pool called Cox's Landing and hooked a 25-pounder. "Sherwood got in at 9:30 p.m., best day he ever had," Peterson wrote in the log. (By this time the enthusiastic Sherwood was calling every day his best.) "Brought in four lovely salmon — two 10 pounds, one 11 and one 12." The next afternoon, after many fish in the morning, Sherwood hooked four salmon "between the pine and spruce trees along the ledges" at the Home Pool. (In my mind, I can see the very place where he hooked those fish.) That day, Peterson and Harris went to town and shipped 12 fish off to their friends by train.

The next day was overcast with a strong west wind, which made casting tough, but these two anglers were still having the trip of their lives. Peterson wrote: "It is now 8 p.m. Parker [a guide] left after a drink and before I get too many I must write of the big day. Already recorded my 3 this morning and got 6 this afternoon, 2 directly opposite Mann's house, and 4 in Upper Cleveland across from the old birch between the pine and spruce. A truly great day for a beginner but holy smoke how I had to reel as they mostly took upstream when hooked. ESS just came in with most lovely fish. He killed six salmon this afternoon at Cox's Landing. Total of 18 salmon, perhaps a record for this camp."

By June 27, Sherwood had his weekly limit of 30 fish and did not cast, but Peterson fished hard, hooking one in Upper Cleveland in the morning and two at Cox's Landing in the afternoon. Peterson wrote: "This ends a most wonderful six days. It is only shadowed by a constant stream of big bombers flying east. More than 25 flew over today and it keeps this horrible war constantly in front of us. But this is a perfect river and killing salmon is one of the greatest sports of all time."

*

Not every year, even in the years of the great fish runs, was like June of 1942. The records break off and then resume in 1951. Weather conditions create the only problems. In 1953, there were heavy rains and high water conditions. By July 3, the trains from Matapédia had carried out 800 boxes of salmon, as opposed to 1,700 at that time the year before. On August 2, 1953, the log reads: "General conditions this year worst in the history of man. Where have we heard that before?"

On Sunday, July 14, 1957, brothers Harold, Arthur and Sydney Howe and two of their friends arrived by car from Squam Lake, New Hampshire. Their guides were Wally Harris, Hudson Deameaux and Clifford Farrer. In their first five days on the river they caught 25 salmon. On the fifth day, the log reads: "In the evening at Cox's, Harold, while wading, hooked a very large salmon and played him 30 minutes, watched by passengers of 3 or 4 cars on the highway and by boys on right bank of river. In attempting to save the fish Harold crossed the river (swimming!) and followed downstream toward the island at the foot of the still water — the salmon then raced on back to the salt water and left angler frustrated and wet through." At the end of six days they had landed 34 fish.

There were many fish in the river, but, the log notes, no blackflies or mosquitoes. The guides told them this was a result of spraying for spruce budworm. What the Howe brothers did not know was that upstream in the headwaters, thousands of young salmon were dying from DDT poisoning. In 1958, the guests who stayed at the camp from July 18 to 25 killed eight salmon, all of which bore marks from nets. Even this far upriver there were signs that salmon were being slaughtered by the tens of thousands in the river estuary and in the ocean.

The club eventually disbanded and the log book lay forgotten in a cupboard at the camp. Arthur Howe's family continues to visit the camp to this day, and stories of the river continue to be told among members of my family, who, like members of the club, have grown to love the waters of the sweet little Upsalquitch.

*

Canadian philosopher George Grant, in his book *Technology and Empire*, writes about the importance of loving our particular roots: "In human life there must always be place for love of the good and love for one's own. Love of the good is man's highest end, but it is the nature of things that we come to know and to love what is good by first meeting it in that which is our own — this particular body, this family, these friends, this woman, this part of the world, this set of traditions, this country, this civilization."

Montana writer Norman Maclean came to know and love what is good by meeting it in his river and in the life of his family. Maclean's autobiographical book *A River Runs Through It* hits home with me. Like Maclean, I am the son of a Presbyterian minister who is a fly fisherman. Like Maclean, I have a younger brother who in many ways reminds me of Maclean's brother Paul, a man of intelligence and great passion who is described simply as "beautiful" and "a fine fisherman." The Maclean boys grew up believing that the fishermen on the Sea of Galilee were fly fishermen "and that John, the favourite, was a dry fly fisherman." The Reverend Maclean informed them that Izaak Walton, the author of *The Compleat Angler*, was not a respectable writer, because "he was an Episcopalian and a bait fisherman." The Reverend Maclean taught his sons that "in the beginning was the Word" and that water runs over the words.

At the end of his story, Norman Maclean ends up as an old man in the half-light of the canyon on the Big Blackfoot River, reflecting on his life. Most of his loved ones are gone. He writes: "Eventually, all things merge into one, and a river runs through it. The river was cut by the world's great flood and runs over rocks from the basement of time. On some of the rocks are timeless raindrops. Under the rocks are the words, and some of the words are theirs. I am haunted by waters."

In his critical work, *Where the Blue Bird Sings to the Lemonade Springs*, American author Wallace Stegner responds to Maclean's words by asking: "Fisherman or not, who is not haunted by waters?" In the final pages of *A River Runs Through It*, Maclean's brother Paul dies a violent death. Stegner writes: "Bluntly, brutally, in a few hundred words, the important part of the story is ended with Paul's life; the shadow falls suddenly on a tale that has been often sunny, even farcical. Time comes down like a curtain, what has been vibrantly alive is only remembered, we are left hollow with loss, and we end in meditation on the Big Blackfoot in the cool of the evening, in the Arctic half-light of the canyon, haunted by waters."

Stegner concludes that Maclean's book is "an invitation to memory and the pondering of our lives." In words far less eloquent and profound than Norman Maclean's, that is what I have been inviting my readers to do in these pages.

*

On my final day at the Upsalquitch camp, I visited Clifford Farrer, one of the guides during the Howe brothers' adventure of 1957. Farrer lives on the farm where he was born, in the river valley right across the field from Hatch's farm. You can see his house when you step out the back door of the Upsalquitch camp. His father, William Guy Farrer, who was a supervisor for the Restigouche Riparian Association, cleared the land on the farm with an axe and a cross-cut saw.

As a child, Farrer says, he could catch salmon anywhere. The Upsalquitch boys would take trout hooks and put hen feathers on them and catch fish. They could snare salmon with a wire loop on a stick as the fish lay "just packed on top of one another" at the mouth of Meadow Brook. Farrer started guiding when he was 16 years old and worked on the river for 30 years. When he worked as a guide at the Upsalquitch Forks, he poled his guests 10 miles upriver. They lived outdoors in tents and he cooked meals for the sports. In those days, he brought salmon up from the river in a wheelbarrow and pushed them to the icehouse at the local store. There, the salmon were packed in wooden boxes with ice and sawdust and shipped south on the train. Many of the fish stayed behind in the icehouse, and people could pick up a salmon at the store anytime they wanted.

Farrer started looking after the old Upsalquitch camp in the 1950s, after the camp manager, his older brother Hedley, a Second World War veteran, drowned in the river. Farrer, who is fine man and a magnificent fly fisherman, has become a great friend of my father over the years. They fish together every August and share a love of the river. "There's nothing as peaceful as being on the water," Farrer says.

After I returned to the camp from Farrer's farm, I visited the Home Pool again and then sat alone on the screen porch and watched the river. When darkness fell on the valley, I stayed where I was for a long time and listened to the water flow past the camp.

*

When Maritime poet Eric Trethewey dreams of rivers, they become rivers of memory and experience. In his poem "Dreaming of Rivers," he writes about holding on to his river dreams before they slip away in the morning light.

Fishing gets better
and you bring the boys in from the dream,
give names to them,
keep on climbing in your mind
toward the thin source of the stream,
push on until, short of it,
deep in the shadow of spruce
and straggle of swale
memory stammers, repeats itself,
and for all your cunning
this river fades into other rivers,
all rivers.

Eventually, Trethewey's dream turns and begins to fall away toward the sea.

And one final shoot through a meadow
to the highway bridge
and beyond where you've been yet,
ocean bound now,
with all those other rivers
blinking silver through the trees, golden in the sun
and though you've lost it again
you know it doesn't end here.

I've been dreaming of rivers since my father and I turned right at the Robinsonville bridge. And though I may lose it from time to time, I know it doesn't end here.

Bibliography

Anderson, E.N. *Ecologies of the Heart: Emotion, Belief and the Environment*. New York: Oxford University Press, 1996.

Bielak, Alex T. and Bruce Tufts. "Late Fall Fishing: Fresh Biological Insights." *Atlantic Salmon Journal* (Autumn 1994) 16-19.

Bielak, Alex T. and Tufts, Bruce. "Late Season Survival." *Fly Rod and Reel* (April 1995) 52-53.

Bodo, Peter. "Reflections of the Sweet Little Upsalquitch." *Atlantic Salmon Journal* (Autumn 1992) 14-16.

Bodo, Peter. "Is a New Age Dawning Among Native Peoples?" *Atlantic Salmon Journal* (Spring 1995) 16-20.

Brislain, Michael. *Bugging the Atlantic Salmon*. Fredericton: Goose Lane, 1995.

Brubaker, Elizabeth. *Property Rights in the Defence of Nature*. Toronto: Earthscan, 1995.

Bruce, Harry. "Bless the Ban. Boost the Buyout." *Atlantic Salmon Journal* (Summer 1992) 14-17.

Bruce, Harry. "Salmo Salar: Keeper of the Peace." *Atlantic Salmon Journal* (Summer 1995) 16-19.

Bruce, Harry and Bernard Tomlinson. "The Governor General Could Have Been a Gillie." *Atlantic Salmon Journal* (Spring 1994) 41-43.

Buchan, John (Lord Tweedsmuir). *Pilgrim's Way*. Cambridge MA: Houghton Mifflin, 1940.

Campbell, J.A. "The Grande Cascapedia." *Atlantic Salmon Journal* (Summer 1983) 20-22.

Carson, Rachel. *The Sea Around Us*. New York: Oxford University Press, 1991.

Chesterton, Gilbert K. *What's Wrong with the World*. London: Cassell, 1910.

Clarke, George Frederick. *Six Salmon Rivers and Another*. Fredericton: Brunswick Press, 1960.

Clarke, George Frederick. *Someone Before Us: Our Maritime Indians*. Fredericton: Brunswick Press, 1968.

Christie, Bonnie. "Rendezvous at Restigouche." *Atlantic Salmon Journal* (February 1982) 19-22.

Curtis, Herb. *The Lone Angler*. Fredericton: Goose Lane, 1993.

Dunfield, R.W. *The Atlantic Salmon in the History of North America*. Ottawa: Department of Fisheries and Oceans, 1985.

Fenety, Jack T.H. "A Tribute to Wilfred Carter." *Atlantic Salmon Journal* (Winter 1988) 9-10.

Gillespie, G.J. "Atlantic Salmon Rearing Station in Operation at Mactaquac." *Atlantic Salmon Journal* 4 (1968) 13-16.

Gingras, Sylvain. *A Century of Sport: Hunting and Fishing in Quebec*. St-Raymond QC: Les Éditions Rapides Blanc, 1994.

Grant, George. *Technology and Empire*. Toronto: Anansi, 1969.

Gruenefeld, George. "Zeroing in on ZEC." *Atlantic Salmon Journal* (December 1982) 10-12.

Haig-Brown, Roderick. *The Salmon*. Ottawa: Department of the Environment, 1974.

Harris, Michael. *Rare Ambition: The Crosbies of Newfoundland*. Toronto: Viking, 1992.

Horwood, Harold. "The Foxes of Beachy Cove." In *Dory Loads:Newfoundland Writings and Art.* Ed. Kevin Major. Portugal Cove, NF: Breakwater, 1974.

Kerswill, C.J. and P.F. Elson. "Forest Spraying and Salmon Angling." *Atlantic Salmon Journal* (October 1964) 20-26.

La Forest, Gerard V. *Water Law in Canada: The Atlantic Provinces*. Ottawa: Information Canada, 1973.

Lockhart, John Gibson. *The Life of Sir Walter Scott*. New York: Dutton, 1957.

Lynch, Charles. *Fishing with Simon*. Scarborough ON: Prentice-Hall, 1992.

Maclean, Norman. *A River Runs Through It*. Chicago: University of Chicago Press, 1979.

Matthews, David Ralph. *Controlling Common Property*. Toronto: University of Toronto Press, 1993.

Meggs, Geoff. *Salmon: The Decline of the British Columbia Fishery*. Vancouver: Douglas & McIntyre, 1995

Mills, Derek and Neil Graesser. *The Salmon Rivers of Scotland*. London: Cassell, 1981.

Netboy, Anthony. *The Atlantic Salmon: A Vanishing Species?* Boston: Houghton Mifflin, 1968.

Netboy, Anthony. *The Salmon: Their Fight for Survival*. Boston: Houghton Mifflin, 1973.

Netboy, Anthony. "Prescription for the Atlantic Salmon's Survival." *Atlantic Salmon Journal* 2 (1975) 16-19.

Newman, Peter C. *The Canadian Establishment*. Toronto: McClelland & Stewart, 1975.

Pearse, Peter H. *Turning the Tide: A New Policy For Canada's Pacific Fisheries*. Ottawa: Supply and Services Canada, 1982.

Pearse, Peter H. *Rising to the Challenge: A New Policy for Canada's Freshwater Fisheries*. Canadian Wildlife Federation, 1988.

Pero, Thomas R. "Come Back, Miramichi, Come Back." *Trout* (Spring 1987) 12-25.

Pero, Thomas R. "Voices from the River." *Wild Steelhead and Atlantic Salmon* (Spring 1994) 50-61.

Pero, Thomas R. "Voices from the River, Part II." *Wild Steelhead and Atlantic Salmon* (Summer 1994) 48-60.

Pero, Thomas R. "St. Orri." *Wild Steelhead and Atlantic Salmon* (Winter 1995) 54-59.

Peterson, Allen E. Jr. and Kevin D. Friedland. "River of Dreams." *Wild Steelhead and Atlantic Salmon* (Summer 1994) 41-47.

Power, Geoff. "Warming Rivers." *Atlantic Salmon Journal* (Winter 1990) 40-42.

Raines, Howell. *Fly Fishing Through the Midlife Crisis*. New York: Doubleday, 1994.

Richards, David Adams. *The Coming of Winter*. Ottawa: Oberon, 1974.

Richler, Mordecai. *Solomon Gursky Was Here*. Markham ON: Viking, 1989.

Robart, Rose. "Cascapedia Indian Agreement: Hopeful Sign in Salmon War." *Atlantic Salmon Journal* (February 1982) 23-26.

Samson, Jack. *Lee Wulff*. Portland OR: Frank Amato, 1996.

Shaw, David. *Tragedy of the Salmon: The Scottish Fishery and the 1986 Salmon Act*. East Grinstead UK: Hillfield, 1995.

Stegner, Wallace. *Where the Blue Bird Sings to the Lemonade Springs*. New York: Viking, 1993.

Trethewey, Eric. *Dreaming of Rivers*. Cleveland: Cleveland State University Poetry Center, 1984.

Tufts, Bruce. "Salmon May Soon be in Hot Water." *Atlantic Salmon Journal* (Summer 1995) 40-42.

Walton, Izaak. *The Compleat Angler*. London: Dent, 1970.

Weber, Peter. *Abandoned Seas: Reversing the Decline of the Oceans*. Washington DC: Worldwatch Institute, 1993.

Weber, Peter. "Safeguarding Oceans." *State of the World 1994*. New York: Norton/Worldwatch Institute, 1994.

Williams, Ted. "The Tragic Folly of Mowing Down Forests." *Atlantic Salmon Journal* (Summer 1993) 18-21.

Williamson, Robert. *Salmon Fisheries in Scotland*. Pitlochry, Scotland: Atlantic Salmon Trust, 1991.

Wolfe, Thomas. *Of Time and the River*. New York: Scribner, 1935.

Wulff, Lee. *The Atlantic Salmon*. New York: Barnes, 1958.

Wulff, Lee. "On Releasing Salmon." *Atlantic Salmon Journal* (Winter 1964) 8-9.

Wulff, Lee. "Where Are We Now? The Desperate Struggle to Avert a Major Disaster in Our Atlantic Salmon Resource." *Eastern Woods and Waters* (Spring 1985) 35-37.

Youngson, Alan. *Spring Salmon*. Pitlochry, Scotland: Atlantic Salmon Trust, 1994.

Acknowledgements

I am indebted to many people on both sides of the Atlantic who took the time to tell me their river stories. I want to single out several people who made a significant contribution to this project and were especially helpful and generous with their time. Dr. Wilfred Carter gave me the wisdom of his many years of working for river conservation and helped me to understand the scope of this story. Dr. Alex T. Bielak offered me his scientific expertise and guidance at crucial times during my research. Jack T.H. Fenety presented me with a wealth of knowledge and made me understand the conservation ethic. Renate Bullock and Vincent Swazey, my newfound friends, opened the door to the Miramichi River. Orri Vigfusson was a kind and generous host in Iceland. Dr. Peter Hutchinson made my research in Scotland possible. The Atlantic Salmon Federation offered me free use of its library and photo archives in St. Andrews, and federation staff were most helpful, especially Bill Taylor, Dr. John Anderson and Danny Bird. J.K. Irving flew me to his beautiful lodge on the Restigouche River and allowed me to see the river and the province in ways that were otherwise not possible. I want to thank J.D. Irving for his generous financial support of the book *Home Pool*.

The newspaper series *Watershed Down*, from which this book is adapted, ran for six weeks on the front page of *The Telegraph Journal*. I want to thank Derek Shelley, a talented editor, who carried out the difficult task of getting all the stories in the newspaper. My managing editor and friend Scott Anderson guided this project from beginning to end. The newspaper series and this book would not have been written without the support of Neil Reynolds, the publisher and editor-in-chief of *The Telegraph Journal* and *The Saint John Times Globe*. The inspiration of this fine newspaper editor is behind every word.

Grateful acknowledgement is made for permission to reprint material from the following sources: Michael Brislain, *Bugging the Atlantic Salmon* (Fredericton: Goose Lane, 1995); Elizabeth Brubaker, *Property Rights in the Defence of Nature* (Toronto: Earthscan, 1995); George Frederick Clarke, *Six Salmon Rivers and Another* (Fredericton: Brunswick, 1960) and *Someone Before Us: Our Maritime Indians*

(Fredericton: Brunswick, 1968); Herb Curtis, *The Lone Angler* (Fredericton: Goose Lane, 1993); R.W. Dunfield, *The Atlantic Salmon in the History of North America* (Ottawa: Department of Fisheries and Oceans, 1985, by permission of the Minister of Supply and Services Canada); George Grant, *Technology and Empire* (Toronto: Anansi, 1963, by permission of Stoddart Publishing, Don Mills ON); Michael Harris, *Rare Ambition: The Crosbies of Newfoundland* (Toronto: Viking, 1992, by permission of Penguin Books Canada, Ltd.); Harold Horwood, "The Foxes of Beachy Cove," in *Dory Loads*, ed. Kevin Major (Portugal Cove: Breakwater, 1974, by permission of the author); Norman Maclean, *A River Runs Through It* (Chicago: University of Chicago Press, 1979); David Ralph Matthews, *Controlling Common Property* (Toronto: University of Toronto Press, 1993); Anthony Netboy, *The Atlantic Salmon: A Vanishing Species?* (Boston: Houghton Mifflin, 1968); Anthony Netboy, *The Salmon: Their Fight for Survival* (New York: Houghton Mifflin, 1973); David Adams Richards, *The Coming of Winter* (Ottawa: Oberon, 1974, by permission of Oberon Press); Mordecai Richler, *Solomon Gursky Was Here* (Markham: Viking, 1989, by permission of Penguin Books Canada Ltd.); Jack Samson, *Lee Wulff* (Portland OR: Frank Amato, 1996); David Shaw, *Tragedy of the Salmon: The Scottish Fishery and the 1986 Salmon Act* (East Grinstead UK: Hillfield, 1995); Eric Trethewey, *Dreaming of Rivers* (Cleveland: Cleveland State University Poetry Center, 1984, by permission of the author); Peter Weber, *Abandoned Seas: Reversing the Decline of the Oceans* (Washington DC: Worldwatch Institute, 1993, by permission of W.W. Norton & Company Inc.); Thomas Wolfe, *Of Time and the River* (New York: Scribner, 1935, with the permission of Scribner, a division of Simon & Schuster; copyright 1935 by Charles Scribner's Sons; copyright renewed 1963 by Paul Gitlin Administrator, C.T.A.); Lee Wulff, *The Atlantic Salmon* (New York: Barnes, 1958, by permission of Joan Wulff).

Grateful acknowledgement is made for permission to reprint photographs from the following sources: the Atlantic Salmon Federation, Robert Baker, Dr. Alex T. Bielak, Dr. Wilfred Carter, Ernest Long, David Nickerson, Thomas R. Pero, Gilbert van Ryckevorsel, the *Telegraph Journal*, Orri Vigfusson.

My father, Philip J. Lee, travelled with me to northern New Brunwick on the best days of research. My mother, Roberta Lee, who has been editing my work for many years, read this manuscript.

I write nothing without my wife Nancy, who encourages and inspires me every day.

Index

Y

This book was designed and typeset by Julie Scriver
of Goose Lane Editions in Fredericton.
Colour separations, printing and binding furnished by
Unipress Limited of Fredericton.

The text face is Goudy Oldstyle in digital form by Adobe Systems.
The paper is Horizon 110 lb.